The Physical Properties of Thin Metal Films

T0203565

Electrocomponent Science Monographs
Edited by Donard de Cogan,
University of East Anglia, Norwich, UK
Founding Editor, D.S. Campbell

The Physical Properties of Thin Metal Films

G.P. Zhigal'skii and B.K. Jones

CRC Press
Taylor & Francis Group
Boca Raton London New York

CRC Press is an imprint of the
Taylor & Francis Group, an **informa** business

CRC Press
Taylor & Francis Group
6000 Broken Sound Parkway NW, Suite 300
Boca Raton, FL 33487-2742

First issued in paperback 2019

© 2003 G.P. Zhigal'skii and B.K. Jones
CRC Press is an imprint of Taylor & Francis Group, an Informa business

Typeset in Times by
Integra Software Services Pvt. Ltd, Pondicherry, India

No claim to original U.S. Government works

ISBN-13: 978-0-415-28390-8 (hbk)
ISBN-13: 978-0-367-39513-1 (pbk)

This book contains information obtained from authentic and highly regarded sources. Reasonable efforts have been made to publish reliable data and information, but the author and publisher cannot assume responsibility for the validity of all materials or the consequences of their use. The authors and publishers have attempted to trace the copyright holders of all material reproduced in this publication and apologize to copyright holders if permission to publish in this form has not been obtained. If any copyright material has not been acknowledged please write and let us know so we may rectify in any future reprint.

Except as permitted under U.S. Copyright Law, no part of this book may be reprinted, reproduced, transmitted, or utilized in any form by any electronic, mechanical, or other means, now known or hereafter invented, including photocopying, microfilming, and recording, or in any information storage or retrieval system, without written permission from the publishers.

Trademark Notice: Product or corporate names may be trademarks or registered trademarks, and are used only for identification and explanation without intent to infringe.

Every effort has been made to ensure that the advice and information in this book is true and accurate at the time of going to press. However, neither the publisher nor the authors can accept any legal responsibility or liability for any errors or omissions that may be made. In the case of drug administration, any medical procedure or the use of technical equipment mentioned within this book, you are strongly advised to consult the manufacturer's guidelines.

British Library Cataloguing in Publication Data
A catalogue record for this book is available from the British Library

Library of Congress Cataloging in Publication Data
A catalog record for this book has been requested

**Visit the Taylor & Francis Web site at
http://www.taylorandfrancis.com**

**and the CRC Press Web site at
http://www.crcpress.com**

Contents

Foreword

The impact of developments in electronics since the invention of the transistor never ceases to amaze and has been the subject of extensive analysis by commentators in the scientific, technological and sociological media. As Editor of Electrocomponent Science Monographs I find it astonishing to consider how things have changed since the Series was originally conceived by the late David Campbell. We endeavour to keep up with the field and most of the recent volumes have been consistent with these aims. Just occasionally, there are good reasons to retrace our steps and this volume, *The physical properties of thin metal films* is just such a case. Thin metal films have been around for a long time, indeed, they have been an important part of the electronics revolution. They were essential for the Fairchild implementation of the integrated circuit and have played a key role ever since.

I believe that I am one of many who have worked with thin metal films without a real understanding of their intrinsic or extrinsic properties. Different organisations have their own lore, their conditions that will ensure that metal A will stick on substrate B, or their special arrangement of layers that minimises electromigration effects. It was quite a surprise when numerical simulations of the write-process in a CD-R did not match the experimental results and it was quite some time before it was "discovered" that the physical properties of thin metal films (e.g. electrical resistivity, thermal conductivity) are quite different from the bulk values. So much time and effort might have been saved if the information which is contained in this book had been readily available. There is quite clearly a need for a comprehensive and informative work on the subject. This book is therefore a very valuable addition to the Series.

The act of starting to write a book on the subject of thin metal films must be a daunting task for any author. The field is so large that it is difficult to know where to start, to know what should be included and what should be omitted. With this in mind it is heartening to see the fruits of an Anglo-Russian collaborative authorship. Professor Zhigal'skii has already published a book on the subject in Russian and as the authors point out in their Introduction, this has provided a basis for this excellent up-to-date work which has been adapted for an international readership.

Donard de Cogan
Series Editor

Introduction to the series

Electrocomponent Science Monographs consists of a set of authoritative reviews on subjects related to electronic and electrical component science and technology. Each has been written by an acknowledged expert in the field and represents an easily accessible source of up-to-date information on either single or closely related topics.

In addition to volumes concerned with particular components, this monograph series also includes reviews on associated systems, particularly in connection with VLSI, microwave semiconductor and optoelectronic technologies. The books are designed to be used by final-year undergraduates, postgraduates and engineers working in the field. It is hoped they will prove both interesting and stimulating.

Donard de Cogan
Series Editor

Introduction

Thin films of conducting materials, such as metals, alloys and semiconductors, are used in many areas of science and technology, especially in modern integrated circuit microelectronics which requires high quality thin films for the manufacture of connection layers, resistors and ohmic contacts. Such conducting films are also important for fundamental investigations in physics, radio-physics and physical chemistry.

When the film thickness decreases, the physical properties of the film change so that in thin films new physical properties appear which are not present in large samples of the same material. The criterion that a film is thin is that the film thickness should be comparable in size with, or smaller than, one of the characteristic physical lengths of the process under study: for example, with the electron mean free path in metals or the de Broglie wavelength of carriers in degenerate semiconductors or semi-metals.

The basic electrical properties of thin films; resistivity, surface resistivity, temperature coefficient of resistance, stability, the intrinsic noise and the degree of non-linearity of the current–voltage curve, are described in Chapter 1. A brief description of the various methods of thin film production are given with the emphasis on thermal evaporation-deposition in a vacuum and in an atmosphere of high purity argon. The main differences in the physical properties which thin films possess compared with bulk materials are due to the thinness of the film and the preparation conditions. The influence on the film properties of the different physical and technological conditions during the preparation is described and related to fundamental ideas about the condensation mechanisms of thin films on to a substrate.

In Chapter 2 the conduction in discontinuous, or island structure, films is considered. In such films non-metallic mechanisms of conduction may take place by the transfer of current carriers through the potential barriers between the islands by thermal activation and tunnelling emission mechanisms.

Chapter 3 is devoted to the electrical properties of continuous thin films. Scattering of the current carriers by the surface of the film produces an additional contribution to the resistance and both classical and quantum size effects. The classical size effect is seen in the conductivity of metal films and the quantum size effect in semi-metal films. Fuchs' theory of the size effect is described together with its experimental verification.

A thin film has a large ratio of surface area to volume and therefore the high surface energy may give a significant contribution to the thermo-dynamic potential of the film-substrate system. This results in a decrease of the melting temperature in thin films compared with a bulk specimen. Also new crystallographic structures, absent in bulk metals, may appear in thin metal films.

In thin films deposited on substrates, high internal mechanical stresses appear, sometimes exceeding the elastic limit of the bulk material by several hundred fold. The physical nature of these internal mechanical stresses and the kinetics of the voids which they can produce are considered in Chapter 4. The size of the internal stress is determined by the details of the preparation of the film. These stresses may produce a density of quasi-equilibrium vacancies which can be many orders of magnitude higher than the concentration in bulk metals, sometimes reaching 0.1-1 at.% at room temperature. The physical phenomena associated with these vacancies are considered in Chapters 5 and 6.

Chapter 5 is devoted to the electrical noise in thin films and thin-film metal–metal contacts. The process of creation and annihilation of quasi-equilibrium vacancies in metal films at different kinds of sources and sinks produces random conductivity fluctuations and generates $1/f$ noise. This vacancy model explains many of the observed experimental results. The wide and continuous distribution of activation energies needed to obtain the $1/f$ spectrum over a wide frequency range is explained by the variation of the activation energy due to local micro-stresses. These quasi-equilibrium vacancies also cause the non-linear effects in the conductivity of metal films which are considered in Chapter 6. If a current flows through a film the temperature, and hence its resistance and equilibrium vacancy concentration, increases to give a cubic non-linearity.

An important commercial phenomenon in thin metal films is electro-migration because the momentum in the current density can be high enough to move the atoms. The reliability of modern integrated circuits is determined, to a considerable extent, by the electromigration immunity of the thin-film interconnects. The general description of electromigration in thin metal film conductors is given in Chapter 7 together with the factors that determine the electromigration resistance. The methods for the non-destructive screening of thin-film devices using $1/f$ noise and the non-linearity parameters are described.

This textbook is a translation into English of sections of well-established Russian texts (by G.P. Zhigal'skii) with complete revision and additions to bring the account up to date and appropriate for an international reader-ship. The book is written clearly at a scientific level which enables the complex physical properties particular to thin conducting films to be easily understood. They are described together with the necessary theory, confirming experiments and applications. The authors have taken a practical approach and give the information necessary to understand how to make high quality thin conducting films with small macro-stresses, a low level of $1/f$ noise and small current–voltage nonlinearities.

The book can be recommended to students, engineers and scientists working in the electronics industry and in fields of pure science.

1 Factors that determine the properties of films

1.1 Electro-physical characteristics of conducting films

Let us consider the basic electro-physical characteristics of metal and other conducting films that are important for their practical application.

1.1.1 Electrical resistivity

The specific resistance (resistivity) of materials, ρ, is measured in SI units of $\Omega\,m$, although $\mu\Omega\,cm$ is frequently used as a more practical unit for metals. Sometimes instead of resistivity, conductivity, σ, is used. This is the inverse of resistivity:

$$\sigma = 1/\rho \tag{1.1}$$

Conductivity is measured in units of $\Omega^{-1}\,m^{-1}$.

Bulk conductivity of metals

According to the classical theory of metals, under the action of an electrical field, E, free electrons gain a directed component of drift velocity added to the random speed of their thermal movement. The current carriers are considered as an ideal gas. The resistance is the result of the loss of this extra momentum of the carriers to the lattice by collisions with the thermal oscillations of the crystal lattice atoms (phonons), lattice defects and impurities.

In an isotropic metal the direction of the current density, j, coincides with the direction of the external electrical field. In this case the bulk conductivity, σ_0, is a scalar quantity and the connection between the current and the field is defined by Ohm's law:

$$j = \sigma_0 E \tag{1.2}$$

In the Drude-Lorentz-Sommerfeld theory of the conductivity of metals, for the case of polycrystal or single crystalline metals with cubic symmetry, the following equations hold for the bulk conductivity, σ_0, and the bulk resistivity, ρ_0 [1.1]:

$$\sigma_0 = \frac{1}{\rho_0} = \frac{ne^2 l}{m v_F} \qquad (1.3)$$

where n is the density of the free electrons in the metal, e is the electron charge, l is the mean free path of the conducting electrons, m is the effective mass of an electron and v_F is the average speed of free electrons on the Fermi surface.

For a given metal, ρ_0 is determined by the value of the mean free path, which in turn depends on the structure of the conductor and the concentration of defects and imperfections in it. Pure metals have the lowest values of resistivity, which is determined by lattice thermal scattering (phonons). The presence of impurities distorts the crystalline lattice and results in the scattering of electrons and an increase in resistivity. In real metal conductors there are various irregularities in the periodicity of a lattice: dislocations, vacancies, divacancies, impurities of alien atoms and others. Thus, in a lattice of a real metal, as well as scattering by phonons the electrons undergo other kinds of scattering.

For a small concentration of impurities and defects in a metal the resistivity can be presented as the sum of two terms [1.1]:

$$\rho_0 = \rho_b(T) + \rho_d \qquad (1.4)$$

where $\rho_b(T)$ is the resistivity of the metal with a perfect crystalline lattice because of the phonon scattering, which depends on temperature; ρ_d is the contribution to the resistivity of impurities and defects of the crystal lattice.

Equation 1.4 is a representation of Mattheissen's rule, which says that the different contributions to the resistivity are independent and hence are added together. In this case the scattering of the current carriers by phonons and by all lattice defects are added [1.1].

The resistance of the majority of reasonably pure metals at room temperature ($T_0 \approx 300\,\text{K}$) is caused mainly by the scattering of the conductivity electrons by phonons so that the first term dominates. At some very low temperature the phonon density becomes very small so that the resistance is caused only by the collisions of the electrons with the structural defects of the lattice and impurity atoms, and then the first term may be neglected. The value of this residual resistance can serve as a measure of the metal purity and the perfection of its crystal lattice. Often the resistance of the sample at the temperature of liquid helium ($T \approx 4.2\,\text{K}$), or the ratio of the resistances at 300 K and 4.2 K, is taken as a practical measure.

The values of resistivity for bulk metal conductors at room temperature range from $1.6\,\mu\Omega\,\text{cm}$ (for silver) up to about $10^3\,\mu\Omega\,\text{cm}$ for some alloys. Values of resistivity, together with some of their other electrical properties at a temperature of 300 K, are given for some metals and alloys in Table 1.1 [1.2].

Table 1.1 Electrical properties of some bulk metals and alloys [1.2]

Metal or alloy	Concentration of electrons, 10^{22} cm^{-3}	Carrier mean free path, nm	Resistivity, $\mu\Omega$ cm	TCR, 10^{-4} K^{-1}
Indium	–	–	9	47
Tin	4.48	–	12	44
Cadmium	–	18.1	7.6	42
Lead	13.20	–	21	37
Magnesium	8.60	33.5	4.5	42
Aluminium	18.06	32.9	2.7	42
Silver	8.85	58.0	1.6	40
Gold	5.90	41.8	2.4	38
Copper	8.85	31.5	1.7	43
Nickel	–	69.5	7.3	65
Iron	–	1.07	9.8	60
Titanium	–	–	42	44
Niobium	–	–	18	30
Molybdenum	–	–	5.7	46
Tantalum	–	–	13.5	38
Tungsten	–	15.9	5.5	46
Chromium	–	–	12.9	45
Alloy, Ni 80% Cr 20%	–	–	100.0	1.7
Alloy, Al 1% Si	–	–	2.9–3.1	–
Alloy, Al 1/2% Cu	–	–	2.9–3.2	–
Alloy, Al 1/2% Si 1% Cu	–	–	3.0–3.2	–
Bismuth	2.8×10^{-4}	2000	130	–

Table 1.2 Electrical and mechanical properties of some metal silicides

Silicide	Resistivity, $\mu\Omega$ cm	CTE, 10^{-6} K^{-1}		Young's modulus, GPa
	[1.3, 1.4]	*[1.3]*	*[1.5]*	*[1.5]*
TiSi$_2$	13–20	12.5	14.5	100
TaSi$_2$	35–50	8.8	16.3	110
MoSi$_2$	40–100	8.2	14.7	100
WSi$_2$	26–70	7.0	13.7	120
PdSi	30–35	–	–	–
PtSi	28–35	–	–	–

Layers of silicides and polysilicon

Silicides, which are compounds of silicon with metals, are used for the interconnects between some components in very large scale integrated circuits, VLSICs, for the electrodes of the gates of transistor structures, for the contacts to areas of silicon forming Ohmic contacts or Schottky diodes and for thin-film resistors [1.3]. For these interconnects in VLSICs the silicides of refractory metals (e.g. WSi$_2$, MoSi$_2$, TaSi$_2$, TiSi$_2$) are mainly used, and for the contact layers to silicon the silicides of the semi-precious metals (e.g. PdSi, PtSi) are used. Silicides are characterised by a high resistance to the failure mechanism of electromigration, which will be considered later in Chapter 7, and their high stability during the later fabrication processes, which may include high-temperature processes. However they have the disadvantage of a greater resistivity than the A1 layers. Some electrical and mechanical properties of silicides are given in Table 1.2, together with their resistivity, the linear coefficients of thermal expansion (CTE) and Young's modulus.

Polycrystalline silicon is used as a natural substitute for the A1 layers in VLSICs. The basic advantages in comparison with A1 are its high stability at high temperatures and its chemical inertness. However the resistivity of polysilicon is higher than that of A1 and the silicides. The polysilicon layers are also used as electrodes for the gates of planar and three-dimensional field effect transistors.

1.1.2   *Mattheissen's rule for thin metal films*

In the last section we saw that the resistance of a bulk metal is determined by the electron scattering in the volume of the sample due to phonons and imperfections. In thin metal films the current carriers may undergo additional scattering at the surface of the boundaries of the film. Assuming that all types of electron scattering are independent, the resistivity of a metal film, ρ_f, can be given from the empirical Mattheissen rule as [1.1, 1.6]:

$$\rho_f = \rho_b(T) + \rho_{df}(n_d) + \rho_s(h) \tag{1.5}$$

where ρ_b, ρ_{df} and ρ_s are the contributions to the film resistance caused by the scattering by phonons, imperfections of the crystalline lattice and the boundaries of the film. The first term depends on temperature, the second depends on the defect concentration, n_d, and the third term depends on the film thickness, h. Depending on the details of the preparation of the film, the contribution of the defect term can considerably exceed the other contributions to the total film resistance. Since the surface resistance is effectively in parallel with the two bulk resistance terms, which are effectively in series, the application of the surface term needs care especially when the relative contribution to the bulk and surface terms is changing. If the thickness of the film is comparable with the mean free path of the electrons in the bulk metal then collisions with the surface will become as frequent as collisions in the volume.

1.1.3   *Sheet resistivity*

A conducting film is usually characterised by a parameter called the sheet resistivity, $\rho_\square$, which is related to the resistivity by:

$$\rho_\square = \rho_f/h = Rb/L \tag{1.6}$$

Here R is the resistance of a film sample with thickness, width and length, h, b and L. The last equality is obtained from the equation $R = \rho_f L/hb$. For a film with a thickness h the ratio ρ_f/h can be taken as a constant. Then we have:

$$R = \rho_\square L/b = \rho_\square n_\square \qquad (1.7)$$

where $n_\square = L/b$ is the number of squares with a side b equal to the width of the film. If $n_\square = 1$ then $R = \rho_\square$, the resistance of one square of a film, which has the units of Ohms per square for any scale of linear dimensions. It depends on the thickness of the layer of the conducting material and also on the conditions of the deposition of the film. The parameter $\rho_\square$ characterises the films for their use in the manufacture of elements of integrated microcircuits.

1.1.4 The measurement of bulk and sheet resistivities

To study the electrical properties of films of metal or other conducting materials produced by various technological methods, it is necessary to measure the resistivity. The choice of the details of the method for making films frequently depends on the results of the measurement of resistivity. If a film is thick enough the contribution to the resistance of the scattering at the film surface can be neglected. For metal films with a low impurity content the resistivity should be close to the bulk value. Therefore to improve a technological process the aim is usually to produce films with a resistivity close to that of the bulk metal. However because of the presence of impurities and defects in the structure of the bulk metal and films the resistivity of a film can differ from the resistivity of a pure metal even for sufficiently thick films.

The two-probe method is the simplest method of measuring the resistivity. It consists of the measurement of the resistance of a suitable exact geometrical shape of the film sample. The sample is usually a strip or meander with constant cross-section. For such measurements the resistance of any contacts and the wires to the sample are negligibly small in comparison with the resistance of the film sample. Because of the small magnitude of the resistivity for the majority of metals, the two-probe method fails to produce high accuracy. Therefore the four-probe, or Kelvin, method is generally used for the measurement of the resistivity of thin metal films.

We consider first the measurement of the resistivity of a film using a special shape for the film sample (Figure 1.1). The current, I_0, from an adjustable current source is passed through the film using probes 1 and 4. The potential difference, U_{23}, developed between probes 2 and 3, is measured with a high impedance voltmeter, V. This measurement of the voltage U_{23} in the absence of any current in the electric circuit of the voltmeter V removes the effect of the potential drops across the resistances of probes 2 and 3 to the sample and the current-carrying wires.

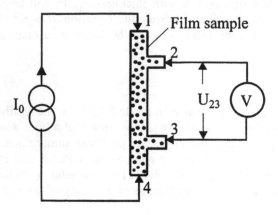

Figure 1.1 Measurement of the resistivity of a film sample using the four-probe method. The current is applied through contacts 1,4 and the potential drop measured across 2,3.

As a rule, the measurements are carried out at a current density, $j \leq 10^5 - 10^6\,A\,cm^{-2}$. The smallest operational current is determined by the sensitivity of the measuring voltmeter and the value of the resistance of the sample between probes 2 and 3. The largest operational current is limited by the allowable heating of a sample. The resistivity of a film is calculated using the formula:

$$\rho_f = (U_{23}/I_0)(hb/L) \tag{1.8a}$$

where U_{23} is the measured voltage between probes 2 and 3, L is the length of the film between probes 2 and 3, and h and b are the thickness and width of the sample respectively.

In the manufacture of integrated microcircuits the measurement of the resistivity of a film is usually carried out on a continuous layer of the metal or other conducting material deposited on a semiconductor wafer with the help of automatic equipment. Tungsten wire probes are usually used. The ends of the probes are sharpened by electrolysis or electro-erosive methods. With a four-terminal method the magnitude of the probe pressure on the contacts does not have much influence on the results of the measurements, however too large a pressure can damage the film. In order to prevent this the ends of the probes are sometimes covered with a layer of indium. The probes 1,2,3,4 (as in Figure 1.1) are arranged on the semiconductor wafer either in a line or at the corners of a square. The voltage between probes 2 and 3 is measured in both directions of flow of the current I_0 and the values U_{23} and U_{32} are recorded. For the calculations the average value of voltage $U_{av} = (U_{23} + U_{32})/2$ is used. This removes any inaccuracies due to a zero error in the voltmeter or to thermoelectric EMFs, which are generated in a circuit composed of different conducting materials if there is a temperature gradient. The surface resistance is determined by the formula:

$$\rho_\square = K U_{av}/I_0 \tag{1.8b}$$

where K is a coefficient defined by the relative position of the probes. K = 4.53 for probes equally spaced in a line and K = 9.09 for probes at the corners of a square [1.7].

The resistivity is calculated using the formula:

$$\rho_f = \rho_\square h = K U_{av} h/I_0 \tag{1.8c}$$

For the measurement of the sheet resistivity using the four probe method the error in the measurement is determined by the type of structure measured and its thickness h. The error can be up to 4–10% for a film thickness of 0.5–2.5 µm. For a smaller measurement error the distance between the probes and the boundary of the sample (the semiconductor wafer) should be greater than $5l$, where l is the distance between the probes, and the value l should considerably exceed the diameter of the point contact of the probe and be more than twice the thickness, h, of the structure.

1.1.5 Temperature coefficient of resistivity

An important parameter of a conducting film is the temperature coefficient of resistivity, α_f, which is determined by the expression:

$$\alpha_f = (1/\rho_f)(d\rho_f/dT) \tag{1.9}$$

The dimension of α_f is the inverse temperature and is expressed in K^{-1} or $°C^{-1}$. The sign of α_f is positive when ρ_f increases with an increase in the temperature.

The magnitude of α_f for a thin film is extremely important for the choice of materials for thin-film elements. For some purposes the temperature coefficient of resistance should be large, for example for bolometers. For other purposes it should be as small as possible, for example for film resistors in microcircuits.

The temperature coefficient of resistivity for bulk metals is always positive. According to classical theory the temperature coefficient of resistance of pure metals in bulk or thin film form should be close to the temperature coefficient of the volume expansion of ideal gases, i.e. close to $1/273\,K^{-1} = 0.00366\,K^{-1}$ because the scattering only depends on the phonon contribution, which is proportional to the absolute temperature. For many metals this value is rather well satisfied as seen in Table 1.1.

The temperature coefficient of resistance of real continuous thin metal films is less than that of bulk metals. This is explained by the greater defect concentration in the film samples and also the influence of the size effect.

That is, there is a greater contribution to the resistivity of the film from the contributions of ρ_{df} and ρ_s which do not depend on temperature.

For a very thin metal film α_f can be negative. The thickness of the layer at which it changes sign depends not only on different metals but also on the conditions under which the samples are produced and the material of the substrate on which the film is deposited. The reason for the occurrence of a negative value of α_f in thin metal films is that the film is not continuous because of the following mechanisms:

1   the film consists of islands separated by vacuum;
2   during condensation the film has absorbed a significant quantity of residual gases. As a result of the interaction of the film material with these residual gases, in particular with oxygen, thin dielectric layers are formed on the grain boundaries. Thus the grains are partly electrically isolated from each other.

In both cases the conductivity mechanisms in the metal films differ from those of the pure metal and a negative temperature coefficient of resistance is found. The explanation of this will be given later in Chapter 2.

To determine the temperature coefficient of resistance it is necessary to take into account the temperature coefficients of linear expansion of the film and substrate. For a free film:

$$\alpha_f = \text{TCR} + \beta_f \tag{1.10}$$

where TCR is the ideal temperature coefficient of the film resistance with no expansion of the lattice and β_f is the coefficient of thermal expansion of the film. For a film strongly bonded to its substrate:

$$\alpha_f = \text{TCR} + \beta_{sub} \tag{1.11}$$

where β_{sub} is the temperature coefficient of linear expansion for the substrate. For pure metals usually $\beta_{sub} \ll \text{TCR}$. For example, for the glass and fused ceramic substrates used in the manufacture of hybrid thick film circuits, $\beta_{sub} \approx 5 \times 10^{-6}\,\text{K}^{-1}$ and for oxidised silicon substrates $\beta_{sub} = (3-4.5) \times 10^{-6}\,\text{K}^{-1}$. These compare with the typical values for TCR near $5 \times 10^{-3}\,\text{K}^{-1}$ given in Table 1.1. Therefore in the majority of practical cases it is possible to assume that:

$$\alpha_f \approx \text{TCR} \tag{1.12}$$

For the experimental determination of α_f we can use the formula:

$$\alpha_f = \left(\frac{(R_1 - R_0)}{R_0}\right)(T_1 - T_0) \tag{1.13}$$

where R_1 and R_0 are the film sample resistances at temperatures T_1 and T_0 respectively, measured over a small temperature interval.

For films of alloys with a small temperature coefficient of resistance and for those with an island structure, which have a large resistivity, the correction for the thermal coefficient of linear expansion can be essential and equations 1.10 and 1.11 are important. At normal operating temperatures the easily melted metals have rather higher values for the coefficient of thermal expansion than do the refractory metals.

For the measurement of α_f, constant temperature enclosures or hot stages are used in the range of temperatures 200–300 °C and cryostats provide cooling of the film sample down to the temperature of liquid nitrogen. The temperature coefficient of resistance measurements at elevated temperatures in air may be accompanied by the oxidation of the film so that they are best carried out in a vacuum. The temperature control can be performed with the help of a thermocouple or similar sensor and a heater. It is necessary to ensure that the thermocouple is well connected to the substrate near the sample or the control temperature may differ from the true temperature of the film. To reveal this undesirable phenomenon the variation of the film resistance with temperature is measured during both the heating and the cooling of the sample. The coincidence of the "direct" and "return" temperature dependencies shows that the measured temperature corresponds to the true temperature of the film.

During the heating of a film irreversible processes can take place, which are connected with the annealing of the non-equilibrium lattice micro-defects. In this case larger changes of resistance are possible than the changes caused by the temperature coefficient of resistance. Figure 1.2 shows the dependence on temperature of the resistance of an aluminium film in vacuum under the following conditions: heating was carried out over 1200 s from room temperature up to 390–410 K; the sample was maintained at this temperature for 1200 s and was then cooled over 1800 s down to 320 K. At a temperature of 390–410 K an irreversible decrease in the film resistance which is caused by the annealing of the crystal lattice point defects, occurs.

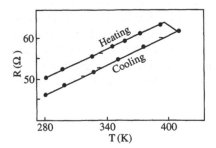

Figure 1.2 Dependence of the Al film resistance on temperature during the first thermocycle.

The linear growth of resistance with temperature is characteristic of metals and occurs because of the scattering of the electrons by phonons [1.1]. The scattering by fixed point defects does not depend on temperature. Therefore, from equation 1.5, stable defects do not influence the size of $d\rho_f/dT$ in equation 1.9 and the introduction, or removal, of additional point defects will only move the straight line of the resistance dependence on temperature and not change its slope. The value of α_f is usually determined for annealed films at rather low temperatures to exclude the irreversible processes connected with the annealing of micro-defects.

1.1.6   The stability of the resistance

The resistivity of a conducting film can change under the influence of electrical current, heat or humidity. The size of this change determines an important characteristic of a film, the stability of resistance. The knowledge of this characteristic is especially important for conducting films used for making thin-film resistors. The relative change of the film resistance under the influence of various factors is defined by the formula:

$$\delta R/R = [(R_2 - R_1)/R_1] \times 100\% \tag{1.14}$$

where R_1 and R_2 are the resistances before and after the test.

The tests are carried out during storage, thermocycling or accelerated ageing. Let us consider what factors define the stability of the resistivity of the film. According to Mattheissen's rule (1.5), the more structural defects there are in a film the higher ρ_f and the smaller the temperature coefficient of resistance (1.9). However the presence of unstable defects and impurities results in an unstable state of a film. In due course relaxation of the structure takes place because of the influence of the environment or the current. The decrease in the number of unstable defects in the film results in a slow decrease in the resistance. Thus there is a contradiction: for a low value of α_f the presence of defects is necessary but unstable defects are not desirable for the stability of the film resistance with time. One of the ways to resolve this contradiction is in the use of thin-film resistors of refractory metals. The defects in these materials have only a small mobility at ordinary operating temperatures. In this case the weak thermal oscillations of the lattice cannot move the imperfections significantly. Therefore the film will be stable in time.

To increase the stability of a film, artificial ageing is carried out by annealing in a vacuum at temperatures of 550–650 K, which considerably exceed the usual operating temperatures.

1.1.7   Internal mechanical stresses

Internal mechanical stresses (macro-stresses) are created in metal films during their formation, processing and storage. These are counterbalanced by the interaction of the film with its substrate. The magnitude of these stresses can reach the strength of the film. They can then result in the spontaneous destruction of the film by a crack or by peeling from its substrate. These mechanical stresses also influence the electro–physical performance of the film. Therefore a study of mechanical stresses in thin films is of interest for the practical application of films and especially for integrated circuits.

The internal stresses are tensile when the film tends to decrease its size in a direction parallel to the surface of the substrate and are then positive in sign. The internal stresses can also be compressive when the film tends to increase its size in the plane of the substrate and then have a negative sign. The strength, or yield point, of thin films exceeds the strength of a bulk material by hundreds of times. The high strength of films is explained by the high density of dislocations, which are strongly fixed by point defects or the surface of the film so that their motion under deformation is hampered. Macro-stresses change under heat treatment or storage. They depend on the film's microstructure, concentration of impurities and the imperfections of the crystalline lattice. The mechanical stresses are described in more detail in Chapter 4.

1.1.8 Electromigration immunity

The electromigration of interconnects in integrated circuits is an important failure mechanism. Electromigration occurs in metal conductors if there is a direct current with a large density, $j > 10^6 \, \text{A cm}^{-2}$. It is the flow of the atoms, both host and impurity, induced by the direct current. Mass transport of the film material normally occurs from the region of the negative contact (cathode) to that of the positive contact (anode). This results in a gradual modification of the electrical properties of conducting films and then to their sudden failure by going open circuit. The subject is described in more detail in Chapter 7.

The electromigration performance of thin-film conductors is usually characterised by the median time of failure. Measurement of the lifetime of interconnects is normally performed by a destructive accelerated stress at high current density and temperature on a large number of samples. For the quality control of thin-film interconnects, and for the separation of good and bad thin-film samples, non-destructive tests using $1/f$ noise, current–voltage non-linearity and multiple electrical measurement are valuable.

Films of pure aluminium have a low electromigration immunity. Thus interconnects in modern integrated circuits are made of Al/Cu alloys, pure Cu or of metal disilicides and polysilicon.

1.1.9 The electrical noise and the non-linearity of the current–voltage characteristics

The electrical noise of a conducting film is the small fluctuations of voltage across the ends of the sample. For thin metal films the major contributions are the thermal noise, a consequence of thermodynamics, and the $1/f$ noise, a property of the particular material of the sample. The latter shows a noise power spectral density, which varies approximately inversely with frequency. This kind of noise is also called excess, flicker or low-frequency (sometimes current) noise. The mechanisms of its origin and physical nature are considered in more detail in Chapters 5 and 6.

For metal films the deviation of the current–voltage characteristic (CVC) from the linear Ohm's law is normally connected with the Joule heating of the film due to the resistive processes of the scattering of the carriers by phonons and vacancies. Other mechanisms are various non-metallic mechanisms of conductivity such as Schottky field emission or tunnelling emission through the thin dielectric layers at grain boundaries. This is discussed further in Chapter 2.

1.2   Outline of thin-film preparation

Metallic, resistive or other conducting films can be prepared by various physical and chemical methods [1.7–1.9]. Most important among the physical methods are thermal evaporation-condensation in a high vacuum or in an atmosphere of an inert gas and ion or ion plasma sputtering. Layers of different semiconductors and some metals can also be prepared by chemical vapour deposition (CVD) [1.7]. The deposition of a film on the surface of an arbitrary material by a chemical reaction occurring in the gas phase near the substrate surface is known as chemical vapour deposition. Layers of compound semiconductors (e.g. GaAs, SiGe) are usually prepared by epitaxial technology and in particular by the molecular-beam epitaxy (MBE) technique [1.8].

The methods of thermal evaporation-condensation and ion sputtering are preferred for metals because of their universality. It should be noted that magnetron sputtering is the main method in the present industrial manufacture of integrated circuits. However it is necessary to take into account the different methods of the film preparation, which are considered in the section below.

The methods of thermal evaporation-condensation and ion sputtering are based on the formation of a beam of atoms (or molecules) of the required materials with their consequent condensation on to the surface of a substrate. These processes are carried out in special vacuum chambers with a very small content of residual gases. The pressure of the residual gases $P_{res} \le 10^{-4} - 10^{-8}$ Pa.

In the process of making the films by vacuum methods three sequential stages can be distinguished:

- evaporation (or sputtering) of the substance in order to make an atomic beam;
- transport of this beam in space from the evaporator to the substrate, or from the target to the substrate in the case of ion sputtering;
- condensation of the atoms on to the substrate with the formation of a film.

1.2.1 Thermal evaporation-condensation in a vacuum

The substance is heated in a high vacuum up to a temperature at which strong evaporation occurs. To obtain a pure film the pressure of the vapour of the evaporated substance should exceed the pressure of the residual gases in the chamber by several orders of magnitude. The resulting beam of vapour spreads in the vacuum space from the evaporator to the substrate. On collision with the surface of the substrate the atoms and molecules of the substance condense on to it to form a film.

For many metals the process of evaporation happens in the sequence: solid phase–liquid phase–vapour phase. However some metals (e.g. chromium, cadmium, magnesium) have a high equilibrium vapour pressure in the solid condition so that they can evaporate directly from the solid phase. This process of evaporation from the solid phase is called sublimation. The process of making films by thermal evaporation-condensation in a vacuum is characterised by the following technological parameters [1.7].

The temperature of evaporation

The vapour pressure of the evaporating substance, and therefore the speed of evaporation, depends on the temperature of the system. The speed of evaporation is determined by the number of atoms evaporating in unit time from unit area of the surface ($cm^{-2} s^{-1}$). The temperature of the substance during the evaporation process is determined by the Joule heating of a direct current through the resistance of the source or with the help of an electron beam hitting the substance itself. In this latter case the electron kinetic energy is dissipated at a spot on the surface of the substance so that its support does not become so hot. This electron beam heat is suitable for the evaporation of refractory metals, for example nickel, tantalum, molybdenum and alloys. In this case a purer and more homogeneous film may be made since there is no direct contact of the hot zone of the evaporated substance with the material of the crucible, which is much less pure.

The intensity of the molecular beam

The intensity of a molecular beam is determined by the number of beam molecules incident upon unit area of the substrate in unit time, v_s, and is expressed in units of $cm^{-2} s^{-1}$. In practice a parameter for the film growth rate (or condensation rate), $w_c = \alpha_c v_s m / D_f$, which is expressed in units of $nm\,s^{-1}$, is usually used, where α_c is the condensation coefficient [1.6] ($\alpha_c \leq 1$), m is the mass of the atom and D_f is the density of the film material, which is usually taken as equal to the density of the bulk material.

The temperature of condensation

The temperature of condensation, T_c, is that of the substrate during the film condensation. To eliminate any molecules that are physically adsorbed on the substrate, mainly molecules of the residual gases and water, it is necessary to increase the temperature of the substrate during the condensation process. In the case of metals the condensation temperature is usually set in the range $T_c = 100$–$200\,°C$. Any further increase in the temperature would reduce the number of physically adsorbed gas molecules but they will become more chemically adsorbed and hence more firmly fixed to the substrate since the chemical adsorption proceeds faster at higher temperatures (see Section 1.3.2).

Degree of vacuum in the chamber

The mean free path of the evaporated atoms, λ_m, depends on the pressure of the residual gases in the chamber. The λ_m parameter is the average distance that the atoms of the evaporated substance move between collisions with the molecules of the residual gases between the evaporator and the substrate. It is usually assumed that the beam atoms have a constant probability of being scattered so that the beam attenuates exponentially. Then, if the pressure of the residual gases is equal to $P_{res} = 1.33\,Pa$, the mean free path of the atoms is $\lambda_m = 4.7\,mm$ and at $P_{res} = 1.33 \times 10^{-4}\,Pa$, $\lambda_m = 47\,m$. The probability of collision of the atoms of the substance with the residual gas molecules in the space between the evaporator and the substrate for modern vacuum installations is very small, a fraction of one percent. The strongest influence of the residual gases on the properties of metal films is by their introduction into the films during the condensation process.

1.2.2 Ion sputter deposition

This method of film deposition is based on the sputtering of the target by bombarding it with fast ions of an operating gas, usually argon. The atoms of the substance, which are expelled from the target surface by the sputtering, spread into the surrounding space and are then deposited on the substrate where they form a film. The process of sputtering is carried out in a gas discharge in which there is a plasma of energetic ionised atoms of the inert gas. There are various types of sputtering arrangements.

Sputtering using a direct current is mainly used for making metal and conducting films. For making dielectric and semiconductor films high frequency, ion plasma, sputtering is used.

Cathode sputtering

In the diode type of apparatus a discharge is formed between two electrodes, the anode and cathode, by applying a high voltage between them in the chamber with argon at a pressure $P_{Ar} \approx 1\text{--}10$ Pa. The gas ions are accelerated in the field near the cathode area and bombard the cathode, which acts as the target. The substrates are placed on the anode where the film of the sputtered material from the cathode is deposited. This method is called cathode sputtering.

The basic disadvantage of cathode sputtering is the low growth rate of the films and, consequently, the contamination of the films by molecules of the reactive gases in the chamber which are mixed with the argon and have a high concentration because of the high argon pressure. On decreasing the argon pressure the discharge dies away, owing to a decrease in the number of electrons produced by ionising collisions with atoms of argon, and the sputtering ceases.

Magnetron sputtering

A variation on cathode sputtering is magnetron sputtering. This method is based on the sputtering of a material by ions of an inert gas (argon) bombarding the target surface. These ions are formed in the plasma of a discharge within a superposition of parallel electric and magnetic fields. If there is a magnetic field the electrons move on helical trajectories. Because of the increase in the length of the path of the electrons passing from cathode to anode the number of ionisation collisions is increased. These support the discharge and allow the sputtering to be carried out at a smaller pressure of argon ($P_{Ar} \approx 10^{-1}-10^{-2}$ Pa). Also, because the plasma is localised on the sputtered surface of the target by the strong magnetic field, the density of the ion current and hence the rate of sputtering is increased. This allows the production of purer films with a smaller content of reactive gases.

Ionic sputtering with negative bias

An effective way of cleaning a film from reactive gases is by placing a negative voltage, U_b, on the substrate during the film preparation. Such a bias encourages the bombardment of the growing film by the positive argon ions and ensures the removal of the gas molecules adsorbed on the film during condensation [1.9]. Thus the removal of impurity atoms from the surface of the growing film, but not the film material itself, will take place if the attractive forces between the metal atoms and the atoms of reactive gases is less than the attractive forces between the metal atoms themselves. For refractory metals (Mo, W, Ta), sputtering with a negative bias is an effective way of removing oxygen and nitrogen atoms. If the negative bias voltage is increased the energy of the ions bombarding the substrate will increase thus improving the cleaning of the film. The resistivity of the refractory metal films becomes nearer to the values characteristic of bulk materials when the potential $U_b = -100$ V [1.7].

1.2.3   Epitaxy of semiconductor thin films

Overview

Although the deposition of metal films is very different from that used for semiconductors, we include a brief discussion for completeness. The term "epitaxy" is applied to the process of making thin single crystal layers on single crystal substrates. The material of the substrate during deposition acts as a nucleating crystal. The epitaxial process differs from the growth of single crystals by the well-known Czochralski method [1.8, 1.9] since the growth of the crystal takes place at a temperature below that of the melt.

The manufacture of modern compound semiconductor devices may require one or more of the epitaxial growth techniques that are available today. These techniques are separated into two basic technologies: liquid phase and vapour phase epitaxy. These techniques will be discussed in this section for GaAs technology. Extremely complex structures are possible using these epitaxial techniques including films composed of GaP, GaAs, GaAlAs, GaAsP, InP, InGaAs and InGaAsP.

In molecular beam epitaxy molecular beams are condensed on to the substrate in a vacuum. If the material of the layer and the substrate are identical, for example, silicon is deposited on silicon, the process is called auto-epitaxy. If the materials of the layer and substrate differ, for example $Al_xGa_{1-x}As$ is grown on GaAs, the process is called hetero-epitaxy. However, in this case the crystalline structure and lattice spacing of the layer and substrate should be similar to grow a single crystal layer.

Substrate requirements

A high quality substrate is needed to obtain the high quality epitaxial structures required for modern solid state devices. For semiconductor epitaxial materials this means a low crystal defect count since substrate defects will propagate through the epitaxial layer to affect the reliability of the final device.

The crystallographic orientation of the substrate is another important consideration. The epitaxial surface morphology, doping levels and growth rates are affected by the substrate orientation and this depends on the epitaxial technique used.

Liquid-phase epitaxy

Liquid-phase epitaxy (LPE) is a single crystal growth process in which material is precipitated from a saturated liquid solution on to a substrate with a crystal structure and lattice spacing similar to that of the material being precipitated.

In liquid phase epitaxy, substrates are usually oriented on axis or very slightly off axis from the (100) or (111) planes. Misorientation of greater than 1° can result in too rapid growth rates and result in surface morphology problems such as terracing. Terracing appears as a ripple effect on the surface and has an adverse effect on subsequent processing. Epitaxial layers grown by LPE are more susceptible to surface defects and irregularities than layers grown using other techniques.

Part of the reason is that LPE involves the saturation of liquid gallium with arsenic by dissolving GaAs in the gallium at an elevated temperature. Epitaxial growth takes place by cooling the entire system down with the saturated melt in contact with the substrate. The surface morphology of the layer is determined by the level of saturation of the melt, the cooling rate, substrate orientation, design of the growth vessel and any oxygen or water vapour contamination in the gas stream. Oxygen is detrimental to LPE growth so that most LPE systems are designed to be operated in a hydrogen ambient, which complicates the safety requirements for this type of system.

Vapour phase epitaxy

Vapour phase epitaxy (VPE) is a crystal growth process in which gases are used to deposit a solid compound on to a crystal substrate with similar composition [1.8].

For vapour phase epitaxial growth, orientations of 2° to 6° off the (100) plane are desirable. Orientations of closer than this can result in rough surface morphology and can affect the stability of the growth.

There are two basic types of VPE systems: trichloride and hydride. The fundamental difference between these is in the arsenic source. While $AsCl_3$, a liquid source, is used in the trichloride system, arsine, a gaseous source, is used in the hydride system.

The trend towards more complex solid state device structures that require submicron layers and selective epitaxy will lead to the introduction of newer technologies such as metal organic chemical vapour deposition (MOCVD) and molecular-beam epitaxy (MBE).

Metal organic chemical vapour deposition

In vapour phase epitaxy the growth of the layer is carried out in a hot reactor, while for MOCVD only the substrate needs to be heated to react the metal organic compounds used as the vapour source. Usually this is done using induction heating or heating by a strong beam of light [1.8]. MOCVD uses vapour sources that involve the pyrolyzation of trimethylgallium or triethylgallium and the source of arsenic is arsine (AsH_4) at approximately 700 °C in a hydrogen atmosphere.

Molecular beam epitaxy

Molecular beam epitaxy (MBE) is a technique that has become increasingly important in the production of the films that are used in very sophisticated compound semiconductor devices [1.8].

The process involves the firing of thermal beams of gallium and arsenic on to a heated substrate under ultra high vacuum conditions; pressures of less than 10^{-8} Pa. The low growth rates used in MBE, of $0.5–5.0\,\mu\text{m}\,\text{hr}^{-1}$, enable precise stoichiometric composition and thickness control and better surface morphology than either vapour phase epitaxy or liquid phase epitaxy. In making the layers of the compound semiconductor GaAs the source of arsenic is contained in a heated crucible and the source of gallium is liquid gallium in another heated crucible. Each crucible is covered by a shutter that can be moved away during the growth cycle to allow the material to be deposited on to the heated substrate while the source is kept at its stable temperature. The shutter feature in MBE systems can produce the sharp doping profile changes which are desirable in many device structures. The disadvantages of MBE are the extremely high initial equipment cost, low wafer throughput and potentially high downtime due to the stringent vacuum requirements.

1.3   The basic film condensation mechanism

1.3.1   The interaction between the vapour and substrate

Initiation of the condensation

Let us consider the state of the metal atom as it goes from the evaporator (source) to the surface of the substrate. The atom leaves the evaporator with an energy kT and, if it moves without collisions with the atoms of the residual gas in the space between the evaporator and the substrate, it keeps this energy up to its impact with the substrate. In the interaction between the atoms of the metal and the substrate, some of the atoms hitting the substrate rebound elastically. Other atoms collide with the substrate inelastically, lose their excess kinetic energy and come into thermal equilibrium with the substrate during a relaxation time, τ_e. The average relaxation time is $\tau_e \lesssim (2\text{–}3)\,\tau_0$, where $\tau_0 \approx 10^{-13}$ s is the period of the lattice thermal oscillations. Thus an atom captured by the substrate loses almost all its kinetic energy during a few periods of oscillation of the lattice atom. The atom becomes adsorbed. Such an atom can move on the surface of the substrate by Brownian motion under thermal diffusion. Some time later, because of thermal fluctuations, the atom may receive a large amount of energy and obtain sufficient velocity to come off the substrate surface and hence re-evaporate. From the moment of adhesion of the atom to the substrate up to its separation the atom moves a mean diffusion path length, λ_s.

Thus, the atoms of metal hitting the surface of a substrate will form on it an adsorbed layer which constantly re-evaporates (sublimes). The rate of sublimation is determined only by the temperature of the substrate and the force of adhesion of the atoms with the substrate surface.

If the rate of influx of atoms on to the substrate is very small, a dynamic equilibrium in the vapour-substrate system is achieved. The flow of re-evaporated atoms from the substrate into the vapour will be equal to the flow of atoms from the vapour on to the substrate. Thus on a substrate a layer of metal is formed, the thickness of which will not increase and consequently the film will not form.

If the number of atoms of metal arriving at the substrate surface is more than the number of atoms re-evaporated from it, then the density of particles in the adsorbed layer will increase and the film will grow. This condition can be reached by an increase in the condensation rate or by a decrease in the temperature of the substrate during the film condensation.

After forming the first monatomic layer the formation of the second layer on to the first adsorbed layer will begin. Now the conditions for the condensation of the metal atoms will vary sharply since the atoms going on to the substrate will be bonded, not with atoms of the substrate surface but with atoms of the same type. In practice, the most frequent case is when the forces of binding between the atoms of the metal are more than the forces of adhesion between the atoms of the metal and the substrate. The latter is usually a dielectric. Therefore the speed of re-evaporation of second and subsequent layers of metal from a substrate will be less than that of the first layer. Then the metal film grows quickly.

1.3.2   The potential energy curve of a condensed atom

Physical adsorption

An evaporated metal atom moving in a vacuum without collisions with the molecules of the residual gases keeps its kinetic energy up to the moment of impact with the substrate. At a close distance from the substrate, about the atomic size, the weak long-range force of attraction acts on the atom. At closer distances a short-range repulsive force acts. The equilibrium position of the atom is where the forces of attraction and repulsion are equal. At this distance the atom will have a minimum potential energy.

Let us clarify the nature of these forces. At first we consider the forces of attraction. Between any two electrically neutral molecules the so-called Van der Waals' force operates. This force take place between atoms or molecules of any elements so that they are universal and are responsible for the condensation of a vapour into a liquid. These forces are weak and are sometimes masked by the stronger forces of interaction, for example the forces of the chemical bonds. Originally this force was introduced to explain the deviation from the ideal of the properties of real gases at high pressure.

The potential energy of interaction of the atom with the substrate surface for these forces can frequently be approximated by an expression of the type $V_{dis} \propto z^{-6}$ where z is the distance of the atom from the substrate.

When the atom is very close to the substrate there is a short-range repulsive force varying exponentially with z. This force is called an exchange or valence force and is the equivalent of Pauli's exclusion principle.

In Figure 1.3, the solid curve shows the dependence of the potential energy of the condensed atom on the distance from the substrate. The potential energy is the sum of the two component: the long-range force and the short-range force. The full curve is sometimes called the adsorption potential and corresponds to the total potential energy of the condensed atom. This curve shows that there is a distance from the substrate, z_m, at which the atom has a minimum potential energy equal to $-V_0$ where the attractive and repulsive forces are equal. The capture of a condensing atom by the substrate corresponds to the atom falling into a potential well with a depth V_0 and then the atom vibrates within it. The binding energy of the atom for physical adsorption is very small and is about a tenth of an electron-volt. The atom will evaporate again if it should gain a kinetic energy equal to, or larger than, this energy that represents the heat of adsorption of the atom by the substrate.

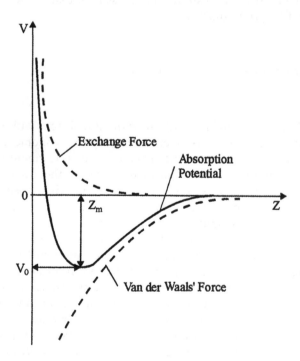

Figure 1.3 The potential energy of an atom condensed on to a substrate.

If the atom adheres to the substrate then one says that it is physically adsorbed and it is called an adsorbed atom. The process of physical adsorption occurs almost instantly. On physical adsorption the atom does not lose its individual properties as there is no interchange of the valence electrons between the atoms of the substrate and the adsorbed atom.

Chemical adsorption

If there is a chemical affinity between the atoms of the film material and the substance of the substrate then, as well as physical adsorption, chemical adsorption is possible and the forces of chemical bonding participate in the adhesion of the atoms. One can distinguish the weak adsorption, determined by homopolar forces, and the strong chemical adsorption, connected with ionic (heteropolar) forces. Quantum mechanics does not give a strict differentiation between the ionic and homopolar forces since they are just the two extreme cases. In the homopolar case the charge is distributed symmetrically between the two interacting atoms as in a hydrogen molecule. If the atoms are not identical, then the symmetry is disturbed. If the symmetry is broken sharply, so that the charge of the bonding electrons is concentrated mainly about one of the atoms, then ionic bonding takes place.

On chemical adsorption the metal atom will form a chemical combination with the atoms of the substrate (for example, an oxide) and the atom loses its individual properties. Frequently chemical adsorption happens after physical adsorption. After staying on a substrate for some time the atom can become chemically adsorbed. In many cases this process requires an activation energy and proceeds much more slowly than physical adsorption but takes place faster at higher temperatures. The energy of the atomic bond with the surface of a substrate during chemical adsorption is about $V_0 \approx 1\,\text{eV}$. A precise separation between physical and chemical adsorption is not always possible. Usually it is considered that if the bonding energy $V_0 < 0.5\,\text{eV}$ then physical adsorption takes place and if $V_0 \geq 0.5\,\text{eV}$ then chemical adsorption takes place.

1.3.3 The lifetime of the adsorbed atom

One of the most important parameters of the condensation process is the average lifetime of the adsorbed atom on the substrate surface. Let us consider the physical meaning of the lifetime. Let the surface of a dielectric substrate be bombarded with metal atoms. As each is adsorbed on the substrate it deposits its excess kinetic energy and the atom participates in thermal motion on its surface with the other atoms.

The adsorbed atom has some finite time of stay on the substrate surface. During this time it interacts with other adsorbed atoms, forming complexes (clusters). If the adsorbed atom does not participate in the formation of clusters, it re-evaporates because it may receive sufficient energy from its thermal motion to come off the substrate.

The lifetime of metal atoms on a substrate surface fluctuates about some average value, τ, which is also called the adhesion time. The lifetime depends on the temperature of the substrate and the strength of the bond between the atoms of the metal and the substrate surface. The formula for the temperature dependence of the lifetime was derived by Frenkel in 1924 [1.10]. The Frenkel formula is:

$$\tau = \tau_0 \exp(V_0/kT) \tag{1.15}$$

where τ_0 is the period of oscillation of the atoms in a direction perpendicular to the surface of the substrate, k is the Bolzmann constant and V_0 is the work necessary to separate the atom from the substrate surface, a latent heat of adsorption.

Equation (1.15) was obtained with the assumption that the number of adsorbed atoms on the surface of a substrate is small compared with that of a full monoatomic layer, so that interaction between the atoms is negligible.

In Table 1.3 the calculated values of the lifetime are given for various adsorption energies at temperatures 300 K and 500 K. It can be seen that changing the temperature by 200 K can change the lifetime by many orders of magnitude.

If the bonding energy of the condensing atom with the substrate surface is small, $V_0 < 0.5\,\mathrm{eV}$ (Van der Waals' bond), after some stay on the surface of the substrate the atom can leave it. If $V_0 > 1\,\mathrm{eV}$ (chemical bond) the adsorbed metal atom can remain on the surface of the substrate practically for ever.

It should be noted that the values of the bonding energies between atoms of different metals and the substrate on adsorption are not known precisely. They depend not only on the nature of the condensing atoms and of the substrate, but also on the condition of the substrate surface and this is very complicated to take into account.

Table 1.3 Calculated values of the adsorbed lifetime for various temperatures and bonding energies

T, K	V_0, eV			
	0.044	0.22	0.88	2.2
300	5×10^{-13}	5×10^{-10}	50	10^{24}
500	3×10^{-10}	1.5×10^{-11}	10^{-4}	10^{9}

1.3.4   Formation of adsorbed complexes and nuclei

In Frenkel's theory of condensation [1.10] the condensing atoms adhere to the surface of the substrate to form a two-dimensional vapour. Then the atoms migrate on the surface, collide and form crystalline nuclei, which grow by gaining new atoms.

The bottom of the potential well (Figure 1.3) for the condensed atom is not flat. Because the atoms of a substrate are located periodically (a crystalline substrate) or quasi-periodically (an amorphous substrate) an additional periodic or quasi-periodic field operates on the adsorbed particles. Figure 1.4 shows the curve of the potential energy, V_s, on the crystalline substrate surface. Due to thermal activation at the surface the adsorbed atom moves on the surface jumping from one potential well to another in a diffusive motion.

For the motion of an atom on the substrate surface a considerably smaller kinetic energy is necessary than for its evaporation. This energy of motion between two adjacent positions of equilibrium is called the activation energy of surface diffusion. The potential barrier has a height V_{s0} but for evaporation the atom must overcome a potential barrier with a greater height, V_0. For comparison the potential barrier appropriating to the energy of evaporation of the adsorbed atom is shown in Figure 1.4 by a dotted line. Therefore adsorbed atoms migrate on the surface from one position of potential energy minimum to another. The magnitudes of the energies V_0 and V_{s0} play an important role in the process of condensation. These magnitudes depend little on the condition of the surface. The exact correlation between them is unknown but it is possible to assume that $V_{s0} \approx 0.2\,V_0$ [1.6] (Table 1.4).

The fact that the activation energy for surface diffusion is rather small is very important for the mechanism of condensation. Experiments to prove the existence of surface diffusion have been performed on both crystalline and amorphous substrates [1.11]. It follows that the atoms hitting a substrate are not fixed at their point of adsorption but migrate on the surface.

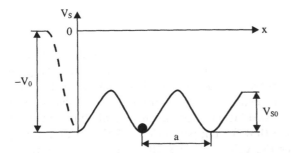

Figure 1.4 The potential energy on the surface of a substrate; x is the coordinate into the surface, a is the lattice constant.

Table 1.4 Experimental data for the energy of bonding and the activation energy for surface diffusion in some systems [1.6]

Metal	Substrate	V_0, eV	V_{s0}, eV
Ag	NaCl	–	0.2
Al	NaCl	0.6	–
Ba	NaCl	3.8	0.65
Cs	W	2.8	0.61
Pt	NaCl	–	0.18
W	W	3.8	0.65
Cu	Glass	0.14	–

The atoms interact between themselves to form stable complexes. Such complexes of atoms are the nuclei for the film condensation. Complexes from two atoms (dimers) will be formed initially. Then other atoms join the dimers through surface diffusion and form a stable nucleus. From these nuclei the grains develop and three-dimensional islands grow. The islands expand to join one another so that a continuous film is formed.

Let us estimate the lifetime of an atom connected with one another in a dimer. Equation 1.15, which calculates the lifetime of an adsorbed atom was deduced with the assumption that the density of the atoms in the layer is small so that the adsorbed atoms interact with the substrate only, instead of with each other. If we modify the energy to separate an atom from the substrate to allow for the interaction of the adsorbed atoms between themselves, that formula will also be true for more dense layers.

Let us call the binding energy between two adsorbed atoms $-\Delta V_1$, where the minus sign means an attractive force. In the physical sense ΔV_1 is the additional work necessary for the separation of the atoms connected in the dimer from the surface of the substrate. Thus the full potential energy of an atom is equal to $-(V_0 + \Delta V_1)$. By analogy with equation 1.15 it is possible to write down an expression for the lifetime of an atom connected in the dimer [1.10]. This assumes that the period of oscillation of the atom does not vary.

$$\tau_1 = \tau_0 \exp[(V_0 + \Delta V_1)/kT] \tag{1.16}$$

Now we compare the lifetime of a single atom, τ, with the lifetime of an atom connected in a dimer, τ_1, for the case of copper deposited on glass at room temperature (T = 300 K). It can be seen from Table 1.4 that in this case $V_0 = 0.14\,\text{eV}$. We will estimate the magnitude of ΔV_1 as the energy of one bond in a metal crystal lattice. For various metals this magnitude is 0.1–0.5 eV. Let us assume that for copper $\Delta V_1 = 0.2\,\text{eV}$. Evaluating equations 1.15 and 1.16 gives the values $\tau \approx 3 \times 10^{-11}\,\text{s}$ and $\tau_1 \approx 8 \times 10^{-8}\,\text{s}$. It follows that the lifetime for an atom connected in a dimer is much greater than the lifetime for an isolated atom. The atoms connected in the dimer are on the substrate for a longer time and interact with other atoms diffusing on the substrate. Therefore dimers act as nucleation centres.

It can be seen from equation 1.16 that the lifetime of a dimer decreases exponentially with increasing temperature so that the number of nucleation centres decreases with an increase in the temperature. Therefore at a high condensation temperature the film is formed with fewer, larger crystallites to give a larger grain film.

1.3.5 The critical condensation temperature

From the study of the deposition of metals on to dielectric substrates it was found that, if the condensation temperature is lower than some magnitude, almost all atoms from a molecular beam adhere to the surface of the substrate. At a higher temperature almost all the atoms are reflected from the substrate. This temperature is called the critical temperature of condensation, T_{cr} [1.12].

In the experiments of Semenov and Hariton [1.12] the critical temperature was determined from the deposition of cadmium on to a copper plate covered with a layer of paraffin. A temperature gradient was created along the plate so that one end was at the temperature of liquid air and the other end was heated with an electrical heater. The temperature of the plate was controlled by thermocouples. The critical temperature was determined by the position of a sharp boundary in the structure of the film. On the part of the plate with a temperature $T < T_{cr}$ a cadmium film was precipitated. On the part of the plate where $T \geq T_{cr}$, the cadmium was not precipitated. It was found that for cadmium on paraffin $T_{cr} \approx 70\,°C$.

Experiments have shown that T_{cr} depends on the condensation velocity of the material. The deposition of metal on to a substrate at a temperature T begins only when the density of the molecular beam achieves a critical magnitude. This fact was first established by Langmuir [1.13] but later experiments clarified the facts. If the speed of the atoms flowing on to the substrate is too small, or the temperature of the substrate is too high, condensation does not take place. This is because, if the lifetime of a single atom on a substrate is too short then there are no collisions between it and the other adsorbed atoms and dimers or other stable complexes do not form.

In order for the condensation to begin it is necessary to increase the intensity of the atom beam on to the substrate at a given temperature of condensation, or to reduce the temperature of the substrate at a given intensity of the atom beam. The critical temperature depends also on the chemical affinity between the materials of the film and substrate.

1.4 Preparation of metal films by thermal evaporation-condensation in an argon atmosphere

In the manufacture of films by thermal evaporation-condensation in an atmosphere of argon, the vacuum chamber is evacuated and then filled with argon of very high purity at the required pressure. Usually the argon pressure in the chamber is $P_{Ar} = 10^{-2}-10^{-1}$ Pa.

Aluminium and chromium films deposited in an atmosphere of argon appear to be more ideal than films deposited in a vacuum. They have a lower level of excess noise, which is a common detector of defects. This is connected with differences in their microstructure. Metal films precipitated in an argon atmosphere have larger grains and their electrical resistivity is close to that of bulk metals. Voids will not form when these films are annealed [1.14, 1.15].

Let us clarify the reasons why films deposited in an argon atmosphere have more ideal properties than films made by thermal evaporation-condensation in a vacuum. The difference from the case of evaporation in a vacuum is that the atoms of the metal evaporated in an atmosphere of an inert gas lose their kinetic energy through collisions with the atoms of the gas in the volume of the chamber, and they join into complexes of two, three and more atoms while still in the vapour phase. At an argon pressure of 10^{-1} Pa the mean free path of the gas molecules is about 6 cm using the exponential law of molecular beam scattering [1.7]. At this pressure, if the distance from the evaporator to the substrate is less than about 15 cm, then fewer than 33% of the evaporated metal atoms reach the substrate without collision with atoms of argon. The evaporated atoms establish thermal equilibrium with the gas in the chamber very rapidly. For example, in the case of aluminium evaporated at a temperature of $T_e = 1500$ K the energy of the atoms will be approximately halved after the first collision. Therefore the majority of the evaporated metal atoms comes into equilibrium with the gas before they reach the substrate. Thus the formation of complexes of metal atoms is possible in the vapour. At first the dimers (complexes consisting of two atoms) will be formed by the triple impact of atoms, one of which is an atom of argon. This carries away the energy and disperses it in the gas medium or to the walls of the chamber. A third and then a fourth atom etc. can join the dimer. Thus primary complexes diffusing in the gas can form into larger aggregates, which then go on to condense on to the substrate. Here they form nucleation centres for condensation. Thus a film with larger grains is formed compared with a film precipitated in a vacuum, when the formation of nuclei of supercritical size occurs by the interaction between single atoms of the metal adsorbed on to the substrate surface [1.6]. The presence of an inert gas in the chamber thus facilitates the formation of atomic complexes and their enlargement in the vapour.

At a high enough pressure of argon the motion of the metal atoms from the evaporator to the substrate can be considered as a continuous stream, while for evaporation in a vacuum the motion of the atoms to the substrate takes place by molecular flow. In the continuous stream the distribution of molecular velocities is a Maxwell distribution. With no external influences the crystallites grow so that the free energy has a minimum value. The decrease of the free energy of a polycrystalline condensate should be accompanied by the improvement of the crystal structure with the formation of the most energetically favourable texture for the film. Such a film will contain fewer imperfections in comparison with the case of condensation from molecular flow [1.7].

1.5 The influence of the deposition parameters on the properties of a film

1.5.1 The film deposition conditions

The actual structure of a metal film depends to a considerable degree on the conditions during its deposition. During the deposition of a film by thermal evaporation-deposition in a vacuum the following parameters are important.

The evaporation temperature

The composition of the particles in a vapour (monoatomic or polyatomic molecules) depends on the evaporation temperature. The composition of the vapour and the energy of the atoms hitting the substrate influence the formation of the amorphous or crystalline microstructure of the film.

The condensation rate

The condensation rate influences the size of the grains and the number of the residual gas molecules introduced into the film. With a higher condensation rate, w_c, metal films are formed with a small-grained structure and with a low concentration of impurities. However at extremely high condensation rates (for aluminium at $w_c \geq 30\,\mathrm{nm\,s^{-1}}$) an increase in the grain size is observed that can be explained by the recrystallisation process (see Section 1.5.3). This is caused by the increase in the substrate temperature produced by the energy of the impinging beam. The increase of the condensation rate also increases the density of vacancies captured within the film, since they have no time to leave the condensate.

The condensation temperature

The condensation temperature is one of the most important parameters that determines the size of the grains and the degree of structural non-equilibrium. The crystals in a film are larger if the adsorbed atoms and their complexes have a higher surface mobility on the substrate during the film condensation. Therefore, well above T_c the crystal size is larger.

The angle of incidence of the molecular beam on the substrate

The angle of incidence of the molecular beam on to the substrate affects the microstructure and the anisotropy of the physical properties, in particular the direction of orientation of the crystals and the internal mechanical stresses in the films.

The physical and chemical properties of the substrate

The physical and chemical properties of the substrate characterise the structural condition of the substrate (amorphous, polycrystalline or monocrystalline), the quality of its surface (roughness, presence of defects) and the chemical affinity between the materials of the substrate and the film. The crystal sizes in a film are larger if the surface is smoother and the chemical affinity between the atoms of the film material and those of the substrate is smaller.

1.5.2 Influence of impurities on the electrical properties of films

The source of the impurity atoms trapped in a condensate are the residual gases in the vacuum chamber, pieces of material already present in the evaporator, the material of the evaporator itself and any contamination on the substrate including any gas molecules adsorbed on it.

Special attention should be paid to the influence of the residual gases. In making films by thermal evaporation in a vacuum, or by ion sputtering, the interaction of the film with any residual gases can happen during both the evaporation of the material and after the completion of the condensation process. In metal films, oxides, nitrides and other chemical combinations form. Thus, the residual gases can influence the chemical composition and the structure of the condensate. The modification of the microstructure in turn can produce a significant influence on the electrical properties of the films. The worse the vacuum, the greater the content of captured gases in the film which produces an increase in the resistivity of such films.

In almost pure films the formation of thin dielectric layers of oxide within the grain boundaries can occur and these grains are then electrically isolated from each other. In this case a non-metallic type of electrical conductivity is observed in the film which is seen as a negative temperature coefficient of resistance (TCR). A negative TCR is observed frequently in films of chromium and the refractory metals which have a high chemical affinity with oxygen.

1.5.3 The effect of heat treatment on the electrical conductivity

Freshly made metal films can exhibit instability in their physical properties during storage with the phase and structural conditions changing. This is connected with the fact that the crystal lattice of the freshly made film contains a large number of micro-imperfections in the structure including non-equilibrium vacancies. Their formation is caused by the enveloping of vacant sites in the crystal lattice of the film by atoms of the molecular beam continuously falling on the substrate during the condensation process. Therefore thin films formed by the deposition of metal vapours can be in a condition of unstable thermodynamic equilibrium, i.e. they have an elevated free energy. The system tends to change into an equilibrium condition with a smaller free energy. Measurements [1.6] of the concentration, n_v, of vacancies in metal films show values of 0.1–1 at.%. The vacancies increase the resistivity of a film by an amount $\Delta\rho_f$. At low vacancy concentrations the increase in the resistivity of a film follows Mattheissen's rule to give a linear increase

$$\Delta\rho_f = A n_v \tag{1.17}$$

where A is a constant of proportionality. For metals $A \approx 1\text{–}3\,\mu\Omega\,cm$ for 1 at.%. Non-equilibrium micro-imperfections in the film structure have specific activation and annealing energies so that heat treatment will remove the imperfections. Thus there is an irreversible decrease in the resistivity of the film. However in some cases on heating the film and keeping it at a defined temperature in a vacuum chamber an increase in the resistivity is observed.

Let us consider in more detail the processes taking place during the heat treatment of films. It is possible to divide the processes into structural, phase and chemical modifications of films.

1 *Annealing of micro-imperfections of various types inside grains.* During this stage of the annealing, processes take place which are connected with a decrease in the concentration of the non-equilibrium point defects inside the crystallites and with a redistribution of the dislocations. Such processes are also accompanied by the formation and removal of small-angle grain boundaries. When such processes take place the resistivity of the film decreases.

2 *Recrystallisation.* During this process the number of randomly oriented crystallites in a film decreases because of changes in the high-angle grain boundaries and the merging of crystallites.

3 *Chemical reactions in the solid phase.* These processes include oxidation in the atmosphere of the residual gases and always increase the resistivity of a film.

Frequently all the above processes take place simultaneously. The probability of this or that process depends on the temperature and duration of the annealing, on the material of the substrate and a series of other factors. It can be assumed that the elimination of imperfections inside grains takes place at a lower annealing temperature than the recrystallisation. Such annealing for $T_{ann} < T_{rec}$, where T_{rec} is the temperature of recrystallisation, i.e. the lowest temperature at which recrystallisation can be observed, is usually called low-temperature annealing. For $T_{ann} \geq T_{rec}$ recrystallisation takes place strongly. Such annealing is called high-temperature annealing. For bulk metals recrystallisation takes place at a significant rate at temperatures much higher than the temperature given by the empirical relation [1.16]:

$$T_{rec} \approx 0.4\, T_{mel} \qquad\qquad (1.18)$$

where T_{mel} is the melting temperature.

For thin metal films the concept of a recrystallisation temperature is arbitrary to some extent, since recrystallisation is frequently observed at lower temperatures than given by 1.18 and can even occur at room temperature. However noticeable recrystallisation in films occurs at high temperatures. For thin films the temperature of recrystallisation depends mainly on their structure and the type of substrate. Despite this, equation 1.18 is frequently used to evaluate the recrystallisation temperature in films. If the recrystallisation process happens at a temperature that is higher than the film condensation temperature, the sizes of the grains after recrystallisation are much smaller than when the film is deposited on a substrate at a temperature $T_c \approx T_{rec}$. This distinction is because the activation energy of the diffusion process at recrystallisation is higher than at condensation and also indicates a higher mobility of the metal atoms on the condensed surface than in the deposited film.

The low-temperature annealing process, connected with the annealing of imperfections inside the grains, occurs prior to the beginning of recrystallisation. If a film is kept at a temperature $T_{ann} < T_{rec}$, then an exponential decrease in the resistance with time occurs because of the annealing of the lattice imperfections and is given by the relation [1.17]:

$$N = N_0 \exp(-t/t_0) \qquad\qquad (1.19)$$

where t is the time of the anneal, t_0 is a constant defining the rate of annealing of imperfections (the characteristic time of annealing), N is the concentration of micro-defects at time t and N_0 is the concentration of micro-defects at $t = 0$.

The resistance of a film on annealing decreases partly irreversibly. At a lower temperature than the annealing temperature the structure of the film should be stable, if the interaction of the film material with any external medium is prevented.

To eliminate the various types of imperfections, different annealing temperatures are needed. If the annealing temperature is increased, the following types of imperfections are eliminated in the order: vacancies, divacancies, complexes of vacancies and dislocations [1.17]. This is explained by the increase in the activation energy for each of these types of imperfections.

With high-temperature annealing the resistance of a thin film sample can be irreversibly increased by recrystallisation, which is accompanied by the growth of crystallites and the increase in the gaps between them. Thus a continuous film can become an island film owing to the formation of breaks between the grains.

The effect of annealing on the resistivity of the film depends also on the material of the substrate. The film resistance decreases more on annealing if the crystal grains of the substrate are smaller and the chemical affinity of the film material to the material of the substrate is greater, that is the energy of adsorption is greater.

To increase the stability of a film, for example for resistors, the temperature of condensation is increased and usually the number of unstable micro-imperfections decreases.

1.5.4 Influence of the film thickness on the conductivity

Very thin metal films usually have a resistivity considerably larger than the resistivity of the bulk metal. This is because the metal conduction in a film starts not at the start of the deposition but at some minimum thickness defined by the conditions of film deposition and the substrate material. In the first stages of metal deposition the film is formed from separate nuclei. The first atoms staying on to a substrate gather in small islands by surface diffusion. Then these increase in size and eventually become so large that they begin to join one another. Then they join together to form a continuous film.

The thickness at which the layer shows metal conductivity is called the critical thickness, h_{cr}. Thus if the film thickness is smaller than the critical thickness the electrical resistance will be much larger up to a value which exceeds the resistivity of the bulk metal by some orders of magnitude. The very high resistivity of super-thin metal films is explained by their discrete structure. Such films are called island films. So for metal films on a dielectric substrate at some thickness, defined by the condensation temperature, a sharp decrease of the film resistance with thickness occurs when the islands join together.

Figure 1.5 shows the typical dependence of the resistivity on thickness for aluminium films deposited by thermal evaporation in a vacuum on to glass substrates at various condensation temperatures ($T_c = 25$, 50 and 120 °C) [1.18]. The critical thickness for aluminium films at these condensation temperatures is approximately 3, 5 and 7 nm respectively.

The critical thickness increases as the condensation temperature increases. This is because the rate of formation of nuclei drops with an increase in the condensation temperature owing to the increase of the mobility of the adsorbed atoms, and the film grows from a smaller number of nuclei. Therefore when these three-dimensional nuclei join together they should reach larger sizes.

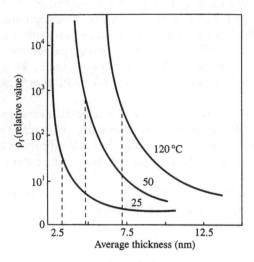

Figure 1.5 Dependence on thickness of the resistivity for aluminium films deposited on glass substrates at various temperatures of condensation [1.18].

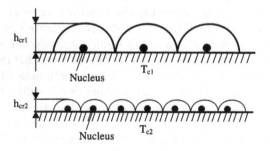

Figure 1.6 Illustration of the decrease in the critical thickness h_{cr} with a decrease of the density of nuclei at different condensation temperatures: $T_{c1} > T_{c2}$.

Figure 1.6 illustrates the increase of h_{cr} with a decrease of the density of nuclei for different condensation temperatures, T_{c1} and T_{c2} ($T_{c1} > T_{c2}$).

For films precipitated on various substrates the critical thickness is very varied. If the precipitated material has a higher affinity with the material of the substrate a continuous film is formed at a smaller thickness. Figure 1.7 shows the dependence on thickness of $\log \rho_f$ for films of gold obtained by thermal evaporation in a vacuum on to various substrates [1.8]. A continuous film for the evaporation on to bismuth oxide is reached at $h_{cr} \approx 5$ nm. If there is a high affinity of the film material with the substrate or there is a low condensation temperature, continuous films may be made with a very

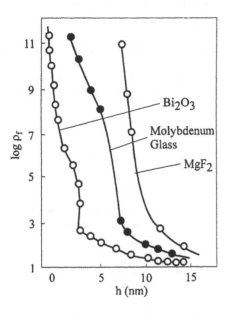

Figure 1.7 The change in the logarithm of the resistivity for gold films on various substrates as a function of thickness for three samples: an oxide of bismuth, a molybdenum glass and MgF_2 [1.19].

small thickness. Thus, for gold on a substrate of bismuth oxide continuous films were obtained by Minn [1.19] with an average thickness of 0.4 nm, which is equivalent to a monoatomic layer. The resistivity of these films had a magnitude of about $10^5 \, \mu\Omega$ cm.

So at different stages of growth the film can be an island film consisting of discrete particles, a network film or a continuous film. Each stage is characterised by its own electrical properties, which are considered in Chapter 2.

Literature for Chapter 1

1.1 Kittel, C., *Introduction to Solid State Physics*, J. Wiley, New York, 1996.
1.2 *Tables of Physical Quantities* (in Russian) Kikoin, I.K. (ed.), Moscow: Atomizdat., Moscow, 1976.
1.3 Murarka, S.P., *Silicides for VLSI Applications*, Academic Press, New York and London, 1983.
1.4 Saraswat, K.L., Brors, B.L. and Fair, I.A., "Properties of low-pressure CVD tungsten silicide for MOS VLSI interconnection", *IEEE Trans. on Electron Dev.*, **ED-30** (1983) 1497–1505.
1.5 Retajczyk, T.F., Sinha, A.K., "Elastic stiffness and expansion coefficient of various refractory silicides and silicon nitride films", *Thin Solid Films*, **70** (1980) 241–247.

1.6   Chopra, K., *Thin Film Phenomena*, J. Wiley, New York, 1969.

1.7   *Handbook of Thin Film Technology*, Maissel, L.I. and Glang, R. (eds), McGraw Hill Book Company, 1970.

1.8   Shur, M., *GaAs Devices and Circuits*, Plenum Press, New York and London, 1987.

1.9   Grosvenor, C.R.M., *Microelectronic Materials*, Adam Hilger, Bristol, 1989.

1.10   Frenkel, Ya.I., "Theorie der Absorption und verwandter Erscheinungen", *Z. Physik*, **26** (1924) 117.

1.11   *Surface Science: Recent progress and perspectives*, **1**, Jayadevaian, T.S. and Vanselow, R. (eds), CRC Press, Cleveland, 1974.

1.12   Semenov, N.N. and Hariton, Yu.B., "To a question about critical temperature of reflection of molecules", *Z. Physik*, **25** (1924) 287.

1.13   Langmuir, J., "The evaporation, condensation and reflection of molecules and the mechanism of adsorbtion", *Phys. Rev.*, **8** (1916) 149.

1.14   Kurov, G.A. and Brilov, I.N., "Influence of argon on some properties of thin aluminium films", *Physics of Semiconductors and Microelectronics*, Ryazan' (in Russian), RRTI Publ. 1979, 75–79.

1.15   Zhigal'skii, G.P., Kurov, G.A. and Siranashvili, I.Sh., "Excess noise and mechanical stress in thin chromium film", *Radiophys. Quantum Electron.*, **26** (1983) 162–168, [*Izv. Vyssh. Uchebn. Zaved. Radiofiz.*, **26** (1983) 207–213 (in Russian)].

1.16   Bochvar, A.A., *Science of Metals* (in Russian), ONTI, Moscow, Leningrad, 1935.

1.17   *Thin Film Microelectronics*, Holland, L. (ed.), Chapman and Hall, London, 1965.

1.18   Shih, D.Y. and Ficalora, P.J., "The effect of hydrogen chemisorption on the conductivity of evaporated aluminum films", *J. Vac. Sci. Technol. A.*, **2** (1984) 225–230.

1.19   Minn, S.S. and Rech, J., *J. Rech. Centre Nat. Rech.*, **51** (1960) 131–160.

2 Conduction in discontinuous films

During the initial stage of growth, when the thickness is less than some critical thickness, a film consists of islands separated from each other by small distances of about 0.1–10 nm. Such a film is called an island or discontinuous film [2.1–2.3]. The electrical properties of such island condensates on a dielectric substrate are very different from the properties of a bulk metal. Their characteristics are closer to the properties of semiconductors. The resistivity of such systems is many orders of magnitude higher than that of bulk metals and is determined mainly by the thickness of the layer. This has been seen in Section 1.5.4. and Figures 1.5 and 1.7.

The conduction in island films is determined basically by the spaces between the islands, through which electrons must jump to allow an electric current to flow. In comparison the resistance of the islands themselves is insignificant.

2.1 Some experimental results

One of the basic characteristics of island films is that their conductivity depends exponentially on temperature. This indicates that the conduction is thermally activated and therefore the temperature coefficient of resistance of island films is negative. The exponential dependence of conduction with temperature was first demonstrated by De Boer and Kraak in a study of very thin layers of molybdenum on glass [2.4]. In Figure 2.1 the typical experimental dependencies of the conductivity, σ_f, for island films of platinum of various thicknesses are reproduced in Arrhenius coordinates [2.2]. These curves are well approximated by the expression:

$$\sigma_f = A_1 \exp(-\psi_a/kT) \tag{2.1}$$

where A_1 is a factor that depends weakly on temperature and ψ_a is the activation energy of conduction, which can be determined from the slope of these curves.

The experimentally observable values of the activation energy for different films lie in the range 0.001–1 eV. As the film thickness decreases the conductivity decreases and the activation energy increases. For the thinner films the islands are smaller and the gaps between them are larger. The experimental values of the activation energy of conduction for gold films as the thickness is reduced are shown in Figure 2.2 [2.3, 2.5]. In these

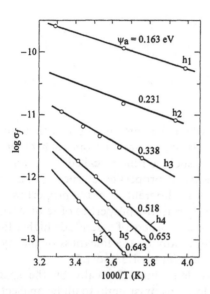

Figure 2.1 Arrhenius plots for platinum films of various thickness with an island structure ($h_1 > h_2 > h_3 > h_4 > h_5 > h_6$) [2.2]. The activation energies of the conductivity calculated from the slopes using equation 2.1 are shown.

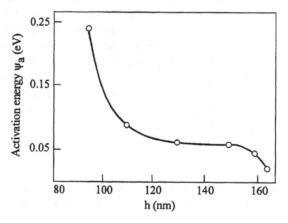

Figure 2.2 Thickness dependence of the activation energy of conduction for gold island films [2.5].

experiments the film thickness is increased by sequential depositions of gold on to the film of the previous thickness under identical conditions of condensation.

Another property of island films is the significant deviation of the current–voltage curves from Ohm's law in strong fields. That is, the conductivity of such films increases with an increase in the field strength. This effect depends on the film structure and the operating temperature. In thicker films the deviation from Ohm's law is observed at smaller voltages. This is because the thicker films consist of larger islands located close to each other. Thus, for any voltage applied to the film, the strength of the field between two adjacent islands will be more than in a thin film containing a greater number of islands. In Figure 2.3 the dependence of the conductivity on the applied voltage for nickel island films of different thickness is shown. The data were taken at 77 K and are plotted in Schottky coordinates $\{\log(\sigma_f/\sigma_f^0), U^{1/2}\}$ where σ_f^0 is the conductivity of the film at the voltage $U = 0$.

As well as these particular properties of island films, adsorbed gases have a strong effect on the conduction and anomalous behaviour of the $1/f$ noise is observed [2.3]. A departure from the normal quadratic dependence of the noise power spectral density on current is observed simultaneously with deviations of the current–voltage curve from Ohm's law. This can be explained by the appearance of non-equilibrium $1/f$ noise, which is further developed in Chapter 6.

These experimental results on the conductivity of island films can be explained by the various mechanisms of conduction, such as thermal electron emission, Schottky field emission and tunnelling emission either directly through the gaps between the islands or through traps in the dielectric substrate.

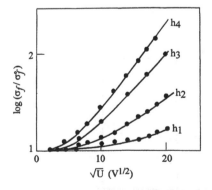

Figure 2.3 The dependence of the logarithm of the reduced conductivity on the square root of the applied voltage for nickel films of various thickness ($h_1 < h_2 < h_3 < h_4$) [2.2].

2.2   The electrical conduction in island films

2.2.1   *The potential barrier between islands*

The basic mechanisms of conduction in island films are related to the existence of a potential barrier between islands. In the case of thermal electron emission and Schottky field emission the electrons pass over a potential barrier when they gain thermal energy which exceeds the height of the barrier. Tunnelling emission is caused by the possibility of an electron, with energy smaller than the height of the potential barrier, passing through the barrier because of quantum mechanical tunnelling. In this case the electrons have an energy close to the Fermi energy of the island [2.1–2.3].

In the roughest approximation the height of the potential barrier at the metal–vacuum boundary is equal to the work function for an electron to leave the metal. For most metals this has a magnitude of 2–5 eV. The potential barrier for this case is shown in Figure 2.4.

The electron energy, W, is measured from the bottom of the conduction band which is taken as zero. Then the energy of all the other conduction electrons is positive. The energy of an electron at rest outside the metal is equal to the height of the potential step, W_0, at the metal–vacuum boundary. This is the energy from the bottom of the conduction band up to the vacuum level. The filled part of the conduction band at $T = 0$ is limited by the Fermi level, F. The *work function*, which is the energy needed for an electron to leave the metal is then given by:

$$\psi_0 = W_0 - F \tag{2.2}$$

Note that the rectangular potential barrier at the metal–vacuum boundary does not correspond to the true profile of the potential near a metal surface. A more accurate picture of the potential distribution can be obtained if the electrical image force is considered. The inclusion of the force is essential for small distances between islands.

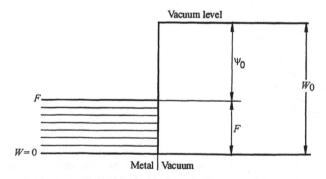

Figure 2.4 A rectangular potential barrier at the metal–vacuum boundary.

A negatively charged electron near the metal induces a distributed positive charge on the metal surface. For a point charge, $-e$, located a distance x_0 above the flat surface of the metal, the metal attracts with a force as though a point charge of the opposite sign, $+e$, exists at a distance x_0 below the surface. The force is localised in the mirror image charge $-e$ (Figure 2.5a). The force of attraction between these charges is equal to:

$$F_{rep} = -\frac{e^2}{4\pi\varepsilon_0(2x_0)^2} \tag{2.3}$$

Thus the potential energy of an electron, defined by this electrical image force is:

$$W_r(x) = W_0 - \frac{e}{16\pi\varepsilon_0 x} \tag{2.4}$$

The curve of the potential energy for the electrical image force at the metal-vacuum boundary with distance from the metal surface is shown in Figure 2.5b. Thus the work function, ψ, of the metal due to the image forces is equal to:

$$\psi(x) = \psi_0 - \frac{e^2}{16\pi\varepsilon_0 x} \tag{2.5}$$

From equation 2.3 it can be seen that the image force has its greatest value very near the surface of the metal and decreases as the electron is moved away. This is a good approximation for distances greater than the lattice constant of the metal. For distances $x < a$ equation 2.3 is not valid because of the influence of the atomic structure of the metal surface. For intermediate distances, $0 \le x < a$, a reasonable approximation is to take the image force constant and equal to the image force at a point $x = a$.

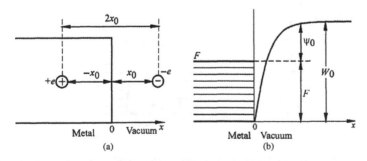

Figure 2.5 (a) Origin of the image force at the metal–vacuum boundary, (b) the potential energy of an electron subject to mirror charge effects.

If the charge $-e$ is located between two parallel metal surfaces a small distance from each other ($d \approx 1$ nm) (Figure 2.6) the electrical image occurs in both electrodes. The potential barrier appears lower, compared with the potential barrier for the case of one electrode (Figure 2.7), at all points between them because of the overlapping of the image forces. The effect of the overlap of the image forces depends on the distance between the electrodes. In Figure 2.7a the mirror-symmetrical curve of the dependence of the electron energies W_{r1} and W_{r2} in the metal–vacuum–metal system is shown for two identical metal electrodes with a gap between the islands $d \approx 10$ nm.

If the metal islands are separated by a distance less than $d \approx 1$ nm the curves merge into one curve $\varphi(x)$ with a maximum φ_0 in the middle (Figure 2.7b). Sometimes this curve is approximated by a quadratic variation [2.3], which is called the parabolic approximation. For comparison in Figure 2.7b the rectangular barrier approximation is also shown. It can be seen that the

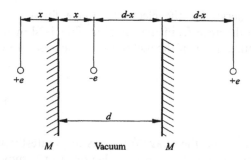

Figure 2.6 Origin of the electrical image forces in the metal–vacuum–metal system.

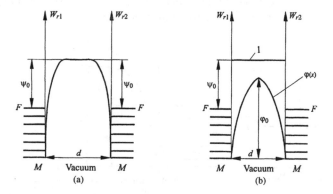

Figure 2.7 The potential energy diagram of the metal–vacuum–metal system, defined by the image forces for various distances d between metallic islands: (a) $d \approx 10$ nm; (b) $d \approx 1$ nm. The approximation of a rectangular barrier is shown with line 1.

effect of the image forces for small gaps between the metal particles is to reduce both the width and height of the potential barrier and to smooth its angles. Such a barrier becomes more transparent to the passage of electrons.

If there is a dielectric, for example a metal oxide or nitride, rather than a vacuum between two islands the potential energy of an electron is determined by the expression:

$$W_r(x) = W_0 - \frac{e^2}{16\pi\ \varepsilon_0\varepsilon\ x} \tag{2.6}$$

where ε is the dielectric permeability of the dielectric.

Figure 2.8 shows the modification of the barrier for different values of the dielectric permeability for values of $\psi_0 = 2\,\mathrm{eV}$ and $d = 2\,\mathrm{nm}$ [2.3, 2.6]. The value of dielectric permeability $\varepsilon = \infty$ corresponds to a rectangular potential barrier with a height ψ_0.

2.2.2 Conduction by thermal electron emission

One of the possible mechanisms of conduction in island films is electron transport between islands by thermal electron emission. This is the emission of electrons from a heated body into a vacuum or the conduction band of a dielectric. It is assumed that any external fields are small and the process of emission does not disturb the local thermodynamic equilibrium inside the body.

Thermal electron emission from metals is possible because, at a temperature $T > 0$, some of the electrons in thermodynamic equilibrium have sufficient energy to leave the metal. Figure 2.9 shows (a) the potential barrier at

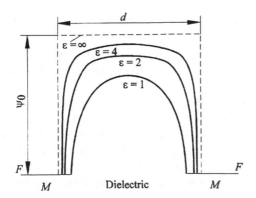

Figure 2.8 The shape of the potential barrier for the metal–dielectric–metal system for various permeabilities, ε, of the dielectric. The dotted line shows the equivalent rectangular potential barrier corresponding to $\varepsilon = \infty$ [2.6].

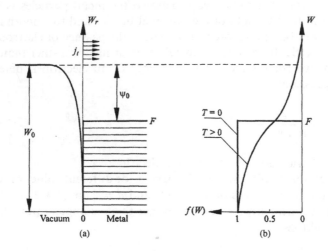

Figure 2.9 Origin of the thermal electron emission current at a metal–vacuum boundary: (a) potential barrier defined by the image forces; (b) Fermi–Dirac distribution function.

the metal–vacuum boundary and (b) the Fermi–Dirac distribution function $f(W)$ for two temperatures. It can be seen that for $T > 0$ there are some electrons with a kinetic energy W greater than the height of the potential barrier ψ_0. Such electrons give rise to the thermal electron emission current, j_T. The saturation current density for thermal electron emission depends on temperature and is given by the Richardson–Dushman formula [2.6]:

$$j_T = A_R T^2 \exp(-\psi_0/kT) \tag{2.7}$$

where A_R is the Richardson constant and ψ_0 is the work function of the metal.

The dependence of the current on temperature is determined mainly by the exponential factor, as the pre-exponential factor $A_R T^2$ depends only weakly on temperature.

For a system of two metal islands with a vacuum space between them, the magnitude of the thermal electron emission current depends exponentially on the height of the potential barrier between the islands. When islands are located very close to each other, the height of the barrier is decreased considerably because of the overlapping potentials of the image forces, as in Figure 2.7b. These arguments give rise to the following expression for the conduction in island films [2.2, 2.3]:

$$\sigma_f = A_1 \exp\left(-\frac{\psi_0 - Be^2/d}{kT}\right) \tag{2.8}$$

where A_1 is a factor depending weakly on temperature and the distance between islands. The term Be^2/d represents the contribution of the image forces to the work function as in equation 2.5 and B is a constant. Thus the effective work function is equal to:

$$\psi_{\text{ef}} = \psi_0 - Be^2/d \qquad (2.9)$$

If the distance d is small enough ($d < 1\,\text{m}$), the effective work function ψ_{ef} can be very small. Calculations show that the decrease of the potential barrier between spherical islands for values typical of their size and for a distance apart of approximately $1\,\text{nm}$ is about $1\,\text{eV}$ [2.3]. However even smaller gaps and barrier heights are possible in island films.

If there is dielectric between the two islands, the potential barrier can be considerably reduced [2.3, 2.6]. In this case, for conduction the electron must go from the metal to the conduction band of the dielectric, the bottom of which is lower than the vacuum level ψ_0 by the electron affinity $\chi = \psi_0 - \psi$, where ψ is an energy from the Fermi level of the metal island up to the bottom of the conduction band of the dielectric as shown in Figure 2.10. The magnitude of the electron affinity is about $1\,\text{eV}$.

2.2.3 Conduction by Schottky field emission

The current due to Schottky field emission is caused by the lowering of the potential barrier height by the external accelerating electric field. The emission of electrons from the metal into the vacuum, or the conduction band of the dielectric, is by thermal excitation over this potential barrier caused by

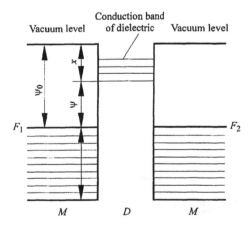

Figure 2.10 The band diagram of the metal–dielectric–metal system for the case when the gap between islands is filled by a dielectric [2.6].

the external electrical field. Let us calculate the Schottky field emission current.

An electric field, E, acts normal to the surface of the metal. For an electron at the surface of the metal a force $F = eE$ acts along the x-axis and tends to pull the electron out of the metal. Along the normal direction this force produces work eEx because the potential energy of an electron decreases by:

$$W_E = -eEx \qquad (2.10)$$

In the absence of an external field the potential energy, W_r, of an electron is determined by the forces of the electrical image as given in equation 2.4. If there is an external field the energy of an electron is the algebraic sum of W_r and W_E and is given by:

$$W(x) = W_0 - eEx - \frac{e^2}{16\pi\varepsilon_0 x} \qquad (2.11)$$

The change in the potential energy of an electron due to the accelerating electrical field is shown in Figure 2.11 by the straight line. The curve W_r shows the potential energy of an electron in the absence of an external field. The curve $W(x) = W_r + W_E$ is the total potential energy of an electron given by equation 2.11. It can be seen that the external accelerating field operates on the potential barrier in two ways. First, it reduces the height of the potential barrier by an amount $\Delta\psi$ and that reduces the work function to $\psi_0 - \Delta\psi$. Second, the external accelerating field reduces the width of the barrier and hence facilitates the tunnelling of electrons through the barrier (see Section 2.2.4).

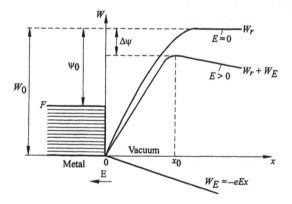

Figure 2.11 Influence of an external electrical field on the potential barrier at the metal–vacuum boundary.

The maximum of the potential barrier is at a distance x_0 from the surface of the metal. This distance is given by the condition:

$$\left(\frac{\mathrm{d}W}{\mathrm{d}x}\right) = -eE + \frac{e^2}{16\pi\varepsilon_0 x_0^2} = 0 \tag{2.12}$$

which gives $x_0 = \sqrt{e/16\pi\varepsilon_0 E}$. By substitution of this value into equation 2.11 we obtain an expression for the height of the potential barrier at the distance x_0 from the surface of the metal:

$$W(x_0) = W_0 - \sqrt{\frac{e^3 E}{4\pi\varepsilon_0}} \tag{2.13}$$

Equation 2.13 shows that if there is an external accelerating field the potential barrier maximum is reduced by about:

$$\Delta\psi = \sqrt{\frac{e^3 E}{4\pi c_0}} \tag{2.14}$$

and the effective work function of electrons from the metal becomes:

$$\psi' = \psi_0 - \sqrt{\frac{e^3 E}{4\pi\varepsilon_0}} \tag{2.15}$$

After substitution of the effective work function magnitude ψ' in the Richardson–Dushman formula (equation 2.7) instead of the ψ_0 magnitude we obtain an expression for the current density due to the emission over the barrier at the metal–vacuum boundary:

$$j_{\mathrm{sh}} = A_R T^2 \exp(-\psi_0/kT)\exp\left(\frac{1}{kT}\sqrt{\frac{e^3 E}{4\pi\varepsilon_0}}\right) = j_T \exp\left(\frac{1}{kT}\sqrt{\frac{e^3 E}{4\pi\varepsilon_0}}\right) \tag{2.16}$$

Here j_T is the saturation current density for thermal electron emission given by equation 2.7. Equation 2.16 was derived first by Schottky and bears his name. The emission of electrons under the action of an electric field is called cold emission or Schottky emission. A characteristic of Schottky emission, as well as of thermal electron emission, is the strong dependence of the current on temperature. A straight line in the coordinates $\{\log(j_{\mathrm{sh}}/T^2), 1/T\}$ shows this dependence. The slope of this increases linearly with $E^{1/2}$.

The band diagram of the metal–vacuum system is shown in Figure 2.12 to illustrate the current flows. Here the flow of electrons that form the thermal electron emission current, j_T, and Schottky emission current, j_{sh}, are shown

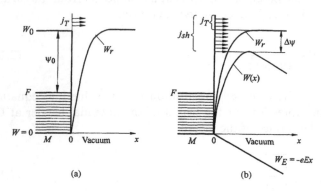

Figure 2.12 Origin of the currents of thermal electron emission and Schottky emission: (a) small electric field; (b) high electric field when the height of the potential barrier is lowered.

schematically by arrows. Figure 2.12a corresponds to the case of a small external field, which does not change the potential barrier height. Figure 2.12b shows the case of a large electric field for which the height and shape of the potential barrier are changed. Figure 2.12 shows that for $T > 0$ a thermal electron emission current is produced by electrons with an energy greater than the work function ψ_0. With an increase in the electric field the height of the potential barrier is reduced and the number of the electrons that can overcome the barrier increases sharply.

If there is an external field then, allowing also for the electric image forces, the effective work function of a metal island is determined by the expression:

$$\psi_{ef} = \psi_0 - \frac{Be^2}{d} - \sqrt{\frac{e^2 E}{4\pi\varepsilon_0}} \tag{2.17}$$

For the conduction through island films it is possible to write the expression [2.1]:

$$\sigma_f = A_1 \exp\left[-\left(\psi_0 - \frac{Be^2}{d} - \sqrt{\frac{e^3 U}{4\pi\varepsilon_0 d}} \right) \middle/ kT \right] \tag{2.18}$$

where $U = Ed$ is the potential drop between islands. Compare this expression with equation 2.8. The Schottky emission reduces to an exponential dependence of the island film conductivity on the inverse temperature and the external electric field.

For weak fields the conduction does not depend on the electric field. But for large electric fields the conductivity varies as $\exp(U^{1/2})$. This can explain the observed experimentally deviation of the current–voltage curve from Ohm's law for island films in strong fields (see Figure 2.3).

2.2.4   Conduction by tunnelling emission

If the conducting islands are separated by gaps from several tenths up to several nanometres then electron transport is possible by tunnelling emission. The carriers are allowed to pass through a potential barrier by the laws of quantum mechanics, even if they have an energy smaller than the height of the potential barrier. For a classical particle such passage is not possible. At low temperatures and for small enough gaps between the islands this mechanism of current flow can dominate. Thus for a gap $d \leq 4$ nm the thermal electron emission current is negligibly small compared with the tunnelling emission current [2.3].

The tunnelling emission current is different from the Schottky field emission current because the electrons pass through a potential barrier whereas in the Schottky field emission current the electrons flow over the barrier. Figure 2.13 shows the band diagram of a metal–vacuum–metal system under the influence of an electric field. Here F_1 and F_2 are the Fermi energies for the two neighbouring islands, which may be different because of the different size of the islands [2.3]. In the figure the approximation of a rectangular potential barrier and the effect of the image forces is shown. The tunnelling emission current, j_{tun}, is due to electrons with energies smaller than the height of the barrier, φ_0. The height and width of the barrier are important here.

Consider the case of a rectangular potential barrier of height φ_1 at low temperatures ($kT \ll \varphi_1$) and for strong electric fields $U = E/d \gg \varphi_1/e$. Here U is the potential between the two islands and E is the electric field

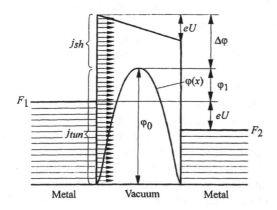

Figure 2.13 Origin of the tunnelling and Schottky emission currents in the metal–vacuum–metal system.

in the gap between islands. The tunnelling current density is described by the Fowler–Nordheim formula [2.1]:

$$j_{\text{tun}} = A_0 E^2 \exp(-B_0 \varphi_1^{3/2}/E) \tag{2.19}$$

where A_0 is a constant depending on the height of the potential barrier.

This conduction mechanism gives an exponential dependence of the current on the magnitude of the inverse electric field, but at weak electrical fields the tunnelling current obeys Ohm's law [2.1].

The tunnelling of electrons can also take place through a dielectric in the gap or the dielectric of the substrate. The possibility of electron tunnelling is determined by the degree of overlap of the electron wave functions in the adjacent islands, which is determined by the size of the Bohr radius, a_B, of the appropriate electronic states. This depends on the dielectric permeability, ε, and the effective mass of the electron, m, in the conduction band of the dielectric and is given by equation [2.3]:

$$a_B = \varepsilon(m_e/m)a_H \tag{2.20}$$

where $a_H = 0.529 \times 10^{-10}$ m is the radius of the first Bohr orbit and $m_e = 9.1 \times 10^{-31}$ kg is the rest mass of an electron.

The effective mass of an electron in a material is usually less than its rest mass. Also the dielectric permeability of a typical dielectric is more than the dielectric permeability of a vacuum. Thus, from equation 2.20, the overlap of the wave functions in a dielectric is greater than in a vacuum ($a_B \gg a_H$), which indicates the dominance of electron tunnelling through a substrate. The conduction through a substrate can take place by the direct tunnelling of electrons or by tunnelling through the fixed energy levels of traps caused by imperfections in the crystal lattice and impurity atoms (Figure 2.14). The type of conduction depends on the shape of the potential barrier and the energy, number and distribution of the levels in the substrate.

If there are positive ions in the dielectric substrate there is an additional increase in the conduction since the height of the potential barrier is lowered (Figure 2.15).

It is possible to explain the non-linearity of the current–voltage curve of discontinuous metal films in strong fields by tunnelling emission. However, tunnelling emission gives a much weaker temperature dependence for the current than is observed experimentally. The exponential dependence of the current on temperature is explained well by thermal electron emission. At the same time the calculated theoretical magnitude of the potential barrier height frequently appears much more than the activation energy of the conductance obtained from experiment. This difference appears even when corrections are made for the overlap of the image forces and the decrease in the potential barrier height in an external electric field [2.3].

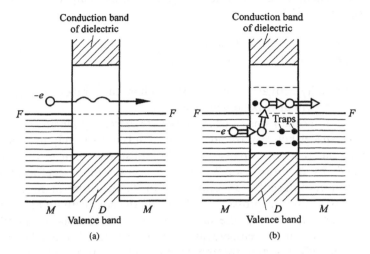

Figure 2.14 The transfer of an electron between metal islands M through a dielectric substrate D by (a) direct tunnelling and (b) through traps in the dielectric substrate.

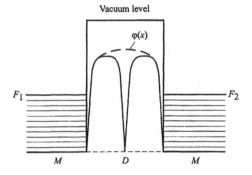

Figure 2.15 Band diagram of a metal–dielectric–metal system in the presence of a positive ion in the dielectric between two metal islands. The dotted line represents the shape of the barrier in the absence of the ion.

2.3 Temperature coefficient of resistance of discontinuous films

In metal island films the temperature coefficient of resistance can have either sign and a range of magnitudes. The behaviour of the temperature coefficient of resistance, as with the conduction, depends on the microscopic details of the film and the properties of the substrate. If the conduction mechanism in a film is due to thermal electron or Schottky field emission then the conduction will have an exponential dependency on temperature,

$\exp(-\psi_{ef}/kT)$. Then, from equations 2.17 and 2.18, the temperature coefficient of resistance is [2.1]:

$$\alpha_f = \frac{1}{\rho_f}\frac{d\rho_f}{dT} = -\frac{1}{\sigma_f}\frac{d\sigma_f}{dT} = -\frac{d\ln\sigma_f}{dT} = -\frac{\psi_{ef}}{kT^2} \tag{2.21}$$

so that for thermally activated conduction in films, the temperature coefficient of resistance will always be negative.

All conduction mechanisms in island films are extremely sensitive to any changes in the spaces between islands. If the islands are strongly bonded to the substrate the thermal expansion of the substrate changes the gaps between islands by $\Delta d/d = \beta_{sub}\Delta T$, where β_{sub} is the linear coefficient of thermal expansion of the substrate and ΔT is the temperature change. Thus the thermal expansion of a substrate will change the conductivity of a film. When the thermal expansion of the substrate is greater than that of the film material, an increase in the temperature will increase the distance between the islands. If this effect is large enough, then the temperature coefficient can become positive. For the majority of substrates the coefficient of thermal expansion is small and its contribution to the final magnitude of the temperature coefficient of resistance can be neglected. However some polymeric substrates, such as teflon (PTFE), have a high coefficient of thermal expansion (up to $10^{-4}\,K^{-1}$), that exceeds that of typical metals by approximately ten times and exceeds that of glass and semiconductor substrates by 20–50 times (see Table 4.2). Therefore the temperature coefficient of resistance of island films deposited on a teflon substrate can appear positive.

The thermal expansion of the substrate results in a modification of the distance between islands and this mainly affects the height of the potential barrier because of the overlap of the electric image forces.

Depending on the dominance of different effects it is possible to observe different sizes of temperature coefficient of resistance, of either sign.

2.4   Percolation conduction in island films

When more material is deposited after the stage when the conducting islands first begin to join, the structure of the film resembles a network. Bridges and gaps between the islands determine the various conduction paths through such films and the total conduction is very sensitive to the number of paths and the physical and chemical changes to the thread-like bridges because of oxidation, annealing and gas adsorption. The diffuse scattering of the current carriers on the surfaces of the small islands and the details of the conduction between the grains largely determines the conduction in network films. The temperature coefficient of resistance of such a film is the sum of the positive metallic temperature coefficient of the particles and the negative temperature coefficient from the gaps between the particles. In the process of filling the network structure with condensate a continuous film is eventually formed and the contribution of the thermally activated mechanisms in the conduction of the film becomes less.

There has been considerable interest in the conduction of such a system just above the percolation threshold, when conduction starts as the amount of material increases. In an ideal system this threshold can be defined by a parameter such as the total coverage of the area. The sudden change in the resistivity constitutes a phase transition and the various parameters approach the critical point with power laws with well defined exponents. There is also a characteristic scale length corresponding to the size of the well conducting clusters which are only slightly connected to each other [2.7].

Many studies have been made of simulations of ideal systems, such as square lattices with resistors connecting the lattice points at random. Experiments have been made on several systems. Of interest here is the observation of non-linearity in the current–voltage characteristics and large $1/f$ noise. The source of these can be either in the physical conduction processes within the fine bridges which dominate the total resistance, or in the dynamic processes of formation and breaking of these bridges near the threshold. It is therefore important to know whether a sample is continuous, granular or near the percolation threshold.

Literature for Chapter 2

2.1 Chopra, K., *Thin Film Phenomena*, J. Wiley, New York, 1969.
2.2 Neugebauer, C.A., "Phenomena of structural imperfections in thin metal films", *Physics of Thin Films*, Hass, G. and Thun, R.E. (eds), Academic Press, New York and London, **2** (1964) 1–64.
2.3 Trusov, L.I. and Holmyansky, V.A., *Island Metal Films* (in Russian), Metallurgiya, Moscow, 1973.
2.4 De Boer, J.H. and Kraak, H.H., *Rec. Trav. Chim. Pays-Bas.*, **55** (1936) 941–948.
2.5 Weitzenkamp, L.A. and Bashara, N.M., "Conduction in discontinuous metal films", *Trans. Metall. Soc. AIME.*, **236** (1966) 351–356.
2.6 Simmons, J.G., *DC Conduction in Thin Films*, Mills and Boon, London, 1971.
2.7 Smilauer, P., "Metal films and percolation theory", *Contemporary Physics*, **32** (1991) 89–102.

3 Size effects in thin films

For continuous metal films the resistance of the grains themselves is the main contribution to the total resistance. The resistance decreases by an order of magnitude in the change from an island structure to a continuous film. If the thickness of a film is comparable with the electron mean free path of the bulk material, the boundaries of the film impose a geometric restriction on the movement of the conduction electrons and, therefore, on the real length of the mean free path of the carriers. This is the so-called classical size effect in the conductivity. This effect produces a decrease in conductivity of a metal film compared with that of the bulk material. Other physical quantities, such as the temperature coefficient of resistance, Hall coefficient and the dependence of the resistivity of the film on a magnetic field, also depend on the thickness of the sample.

The quantum size effect occurs if the thickness of the film is comparable with the de Broglie wavelength of a conduction electron.

A thin film has a large ratio of surface area to volume and therefore possesses a high surface energy that gives a considerable contribution to the thermodynamic potential of the film-substrate system. This results in a decrease of the melting temperature of thin films compared with a bulk specimen. Also new crystallographic structures, absent in bulk metals, appear in thin metal films.

In this section we will consider some aspects of the size effect theories in thin films and describe some experimental results.

3.1 Classical size effects in conductivity

It has been known from Stony's work in 1898 that the conductivity of metal films is less than the conductivity of the bulk material. Thomson put forward the theory of the size effect in conductivity in 1901.

3.1.1 Thomson's formula

The following assumptions were made by Thomson to derive the formula, bearing his name, for the conduction in films [3.1]: (1) the length, l, of the electron mean free path in the volume of the metal is constant and the same for all conduction electrons; (2) after a collision by an electron with the film surface, the probability that it will deflect into any solid angle, $d\omega$, is $d\omega/2\pi$, and thus the probability of scattering does not depend on the initial or final directions of the electron motion. It is also assumed that $l > h$, where h is the thickness of the film. Figure 3.1 shows the model adopted for the derivation of Thomson's formula. The electrons begin their motion from a point A at some distance z_0 from a film surface and move at an angle θ with respect to the z axis ($\theta_1 \leq \theta \leq \theta_0$).

It follows from Figure 3.1 that the actual length of the electrons mean free path in a film, l_f, as a function of the direction of the electron motion is given by:

$$l_f = \begin{cases} \dfrac{h - z_0}{\cos \theta}, & 0 \leq \theta \leq \theta_1; \\ l, & \theta_1 \leq \theta \leq \theta_0; \\ -\dfrac{z_0}{\cos \theta}, & \theta_0 \leq \theta \leq \pi \end{cases} \tag{3.1}$$

where $\cos \theta_1 = (h - z_0)/l$, $\cos \theta_0 = -z_0/l$.

The average length of the mean free path of the conduction electron is calculated from the average, $\bar{l}$, of l_f over all distances and angles:

$$\bar{l} = \frac{1}{h} \int_0^h dz \int_0^\pi l_f \sin \theta d\theta \tag{3.2}$$

After the calculation of the integral in equation 3.2 using equation 3.1 we obtain:

$$\bar{l} = \frac{h}{2} \left(\ln \frac{l}{h} + \frac{3}{2} \right) \tag{3.3}$$

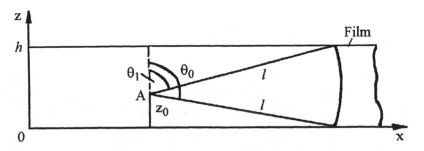

Figure 3.1 The model used in the derivation of Thomson's formula for the size effect in the film conduction.

As the conductivity of a film, σ_f, is proportional to the mean free path it is easy to obtain a formula for the ratio of the thin film conductivity to the bulk conductivity:

$$\frac{\sigma_f}{\sigma_0} = \frac{1}{2}\gamma\left[\ln\left(\frac{1}{\gamma}\right) + \frac{3}{2}\right] \qquad (3.4)$$

where $\gamma = h/l \le 1$ is the normalised thickness.

Equation 3.4 was derived by Thomson. He did not take into account the electrons that started their motion from a surface of the film. To obtain a true value for l it is necessary to take into account all the electrons in the metal at a given moment and to average their free path lengths. Also the statistical distribution of the free path length about the average value was not taken into account. In 1938 Fuchs made a calculation for a model of free electrons with a spherical Fermi surface. Subsequently this theory was applied to the galvanomagnetic effects and was generalised to the cases of arbitrary band structure and the scattering of carriers from two surfaces of a film with different scattering parameters. Fuchs' theory is discussed further here since it is simple and useful for practical applications [3.1, 3.2].

3.1.2 *Fuchs' size effect theory*

The phenomenon of the size effect was analyzed by Fuchs with the help of the distribution functions of the conduction electrons with energy $f(\bar{r}, \bar{p})$, where $\bar{r}, \bar{p}$ are the position and momentum of a particle, which in turn is a solution of the Boltzmann equation.

The Boltzmann equation for the distribution function of the conduction electrons is derived by equating the rate of change of the function to the changes in velocity due to the application of an external field and to collisions. The character of the scattering, diffuse or elastic (specular), is taken into account by the boundary conditions, which are assumed.

If there is both an electric field, $\overline{E}$, and a magnetic field, $\overline{H}$, the Boltzmann equation for quasi-free electrons is written as:

$$-\frac{l}{m}(\overline{E} + \bar{v}\overline{H})\nabla_v f + \bar{v}\nabla_r f = \left(\frac{\partial f}{\partial t}\right)_c \qquad (3.5)$$

Here f is the Fermi–Dirac function of the velocity vector, $\bar{v}$, and the position vector $\bar{r}$. The term $\bar{v}\nabla_r f$, containing the derivatives of the function f on the coordinates, takes into account the heterogeneity in the distribution of the conduction electrons in space that permits the analysis of the size effect.

It is possible in the general case to express the term $(\partial f/\partial t)_c$ through a relaxation time, τ_r, that characterises the relaxation process of the non-equilibrium distribution function, f, to the equilibrium function, f_0, when the external influence is taken away. Thus an approximation for the relaxation time is made:

$$(\partial f / \partial t)_c = -(f - f_0)/\tau_r \qquad (3.6)$$

If v is the average velocity of electrons described by a relaxation time, τ_r, then the mean free path is given by $l = v\tau_r$.

We shall evaluate equation 3.5 for the case of a metal film with a thickness h along the z-axis with an electric field, E, directed along the x-axis. It is convenient to divide the distribution function, f, into two parts: an equilibrium function, f_0, and non-equilibrium one, f_1. Then:

$$f = f_0 + f_1(v, z) \qquad (3.7)$$

and $(\partial f_0 / \partial t) = 0$.

If we neglect the term $\bar{v}\nabla_r f_1$ which constitutes a departure from Ohm's law, then equation 3.5 reduces to:

$$\frac{\partial f_1}{\partial z} + \frac{f_1}{\tau_r v_z} = \frac{eE}{m v_z} \frac{\partial f_0}{\partial v_x} \qquad (3.8)$$

which has a general solution:

$$f_1(v, z) = \frac{eE\tau_r}{m} \frac{\partial f_0(v, z)}{\partial v_x} \left[1 + F(v) \exp\left(-\frac{z}{\tau_r v_z} \right) \right] \qquad (3.9)$$

where $F(v)$ is an arbitrary function of v, defined by the boundary conditions. For the boundary conditions we shall assume that each electron trajectory ends with diffuse scattering after a collision with a surface with the full loss of any drift velocity. Thus the number of electrons leaving each surface of the film in any solid angle, and hence the distribution function, does not depend on the direction. The relaxation process for scattering at the surface is the same as that in the volume of a metal film:

$$f_0(v, z) = f_0(v) \qquad (3.10)$$

where $f_0(v)$ is the Fermi–Dirac distribution function.

Let us now apply the boundary conditions for the calculation of the function $F(v)$. At the surface where $z = 0$ the function $f_0(v, z)$ should not depend on the direction of the velocity v for all $v_z > 0$. Since the function $f_0(v)$ depends on the absolute value of the velocity only, then the first boundary condition $f_1(v, 0) = 0$ gives $v_z > 0$ for all velocities:

$$1 + F(v) = 0 \qquad (v_z > 0) \qquad (3.11)$$

Similarly for $z = h$ we obtain:

$$1 + F(v) \exp\left(-\frac{h}{\tau_r v_z}\right) = 0 \qquad (v_z < 0) \tag{3.12}$$

Thus there are two values of f_1 depending on the direction of the electron from the surface ($v_z > 0$) or towards it ($v_z < 0$):

$$f_1(v, z) = \begin{cases} \dfrac{eE\tau_r}{m} \dfrac{\partial f_0}{\partial v_x}\left[1 - \exp\left(-\dfrac{z}{\tau_r v_z}\right)\right], (v_z > 0) \\[4mm] \dfrac{eE\tau_r}{m} \dfrac{\partial f_0}{\partial v_x}\left[1 - \exp\left(\dfrac{h - z}{\tau_r v_z}\right)\right], (v_z < 0). \end{cases} \tag{3.13}$$

The current density at a point with a coordinate z is:

$$j(z) = -2e\left(\frac{2\pi m}{\hbar}\right)^3 \iiint v_x f_1(v, z)\, dv_x\, dv_y\, dv_z \tag{3.14}$$

where $\hbar$ is Planck's constant.

The evaluation of this integral can be simplified if we put it into polar coordinates v, θ, φ in velocity space, so that:

$$v_z = v \cos\theta = \frac{v}{a}; v_x = v \sin\theta \cos\varphi; v_y = v \sin\theta.$$

Then we have for the current density:

$$j(z) = -2e\left(\frac{2\pi m}{\hbar}\right)^3 \iiint v \sin\theta \cos\varphi f(v, z) v^2 \sin\theta\, d\theta d\varphi dv \tag{3.14a}$$

The average current density in a film, j_f, is obtained by the average of the current over all z from 0 up to $z = h$:

$$J_f = <j> = \frac{1}{h}\int_0^h j(z)\, dz \tag{3.15}$$

The equation for the current density in the bulk metal, j_0, is obtained if we assume, after the integration of equation 3.15, that $h = \infty$. The effective conductivity according to equation 1.2 is $\sigma = j/E$ so that:

$$\frac{\sigma_f}{\sigma_0} = \frac{j_f}{j_0} \tag{3.16}$$

After calculation of the integrals in equations 3.14a and 3.15 it is possible to obtain an equation for the relative conductivity of the film:

$$\frac{\sigma_0}{\sigma_f} = \frac{\rho_f}{\rho_0} = \frac{\varphi(\gamma)}{\gamma} \tag{3.17}$$

where $\gamma = h/l$ and the function $\varphi(\gamma)$ is determined by the relation:

$$\frac{1}{\varphi(\gamma)} = \frac{1}{\gamma} - \frac{3}{8\gamma^2} + \frac{3}{2\gamma^2} \int_1^\infty \left(\frac{1}{a^3} - \frac{1}{a^5} \right) e^{-\gamma a}\, da \tag{3.18}$$

where $a = 1/\cos\theta$.

In the extreme cases of very thin films ($\gamma \ll 1$) and for very thick films ($\gamma \gg 1$) equation 3.17 gives:

$$\frac{\sigma_f}{\sigma_0} \approx \frac{3}{4}\gamma \left[\ln(1/\gamma) + 0.42\right], (\gamma \ll 1); \tag{3.19}$$

$$\frac{\sigma_f}{\sigma_0} \approx 1 - \frac{3}{8\gamma}, (\gamma \gg 1). \tag{3.20}$$

Equation 3.20 is exact for $\gamma \gg 1$ although it is a very good approximation for $\gamma \geq 1$.

In equations 3.19 and 3.20 it was assumed that the scattering at the film surface is completely diffuse. The properties of the scattering are determined by the roughness of the surface, the presence of impurities on the surface, the boundary condition at the surface of the wave functions of the current carriers and also other factors.

The above conclusion can be extended to the case when a fraction of electrons, P (the mirror parameter), is scattered elastically at both surfaces of the film with only a change in the sign of the velocity component normal to the surface, v_z, and the remaining fraction of the electrons $(1-P)$ is reflected diffusely with a full loss of their drift velocity. For elastic impacts the kinetic energy of an electron is conserved but for diffuse scattering it is not.

Equation 3.9 is the usual equation for the distribution function of the electrons in an electrical field. The boundary conditions to determine the function $F(v)$ are different for each particular case. We shall consider the case when the mirror parameter has a value to represent the rough surface approximation. From the symmetry of the problem it follows that $f(v_x, v_y, v_z, z) = f(v_x, v_y, -v_z, h-z)$.

Therefore it is possible to alter equation 3.9 to:

$$f_1(v, z) = \begin{cases} \dfrac{eE\tau_r}{m}\dfrac{\partial f_0}{\partial v_x}\left[1 + F(v)\exp\left(-\dfrac{z}{\tau_r v_z}\right)\right] = f_1(v), (v_z > 0); \\[3mm] \dfrac{eE\tau_r}{m}\dfrac{\partial f_0}{\partial v_x}\left[1 + F(v)\exp\left(\dfrac{h-z}{\tau_r v_z}\right)\right] = f_2(v), (v_z < 0). \end{cases} \tag{3.21}$$

The distribution function of the electrons reaching the surface at $z = 0$ is $f_0(v) + f_2(v, 0)$.

The distribution function for the fraction, P, of the elastically scattered electrons leaving the surface is given by $P[f_0(v) + f_2(-v, 0)]$. The rest of the electrons are not dispersed elastically and give a contribution, g, which does not depend on the scattering angle. Thus the total distribution function of the electrons leaving the surface is given by:

$$f_0(v) + f_1(v, 0) = P[f_0(v) + f_2(-v, 0)] + g \qquad (3.22)$$

After substitution of $f_1(v, 0)$ from equation 3.21 into equation 3.22 we find that the fraction of the electrons which are dispersed inelastically is:

$$g = f_0(v)(1 - P) + \frac{eE\tau_r}{m} \frac{\partial f_0}{\partial v_x} \left\{ (1 - P) + F(v) \left[1 - P\exp\left(-\frac{h}{\tau_r v_z}\right) \right] \right\} \qquad (3.23)$$

The magnitude of g should not depend on the direction of the velocity and the equation in the outer brackets should be equal to zero. From this we find $F(v)$ and from equation 3.21 we obtain:

$$f_1(v) = \begin{cases} \dfrac{eE\tau_r}{m} \dfrac{\partial f_0}{\partial v_x} \left[1 - \dfrac{1-P}{1 - P\exp(-h/\tau_r v_z)} \exp\left(-\dfrac{z}{\tau_r v_z}\right) \right], (v_z > 0); \\[4mm] \dfrac{eE\tau_r}{m} \dfrac{\partial f_0}{\partial v_x} \left[1 - \dfrac{1-P}{1 - P\exp(-h/\tau_r v_z)} \exp\left(\dfrac{h-z}{\tau_r v_z}\right) \right], (v_z < 0) \end{cases} \qquad (3.24)$$

The effective conductivity is calculated just as in the case of completely diffuse scattering. Thus:

$$\frac{\sigma_0}{\sigma_f} = \frac{\rho_f}{\rho_0} = \frac{\varphi_P(\gamma)}{\gamma} \qquad (3.25)$$

where

$$\frac{1}{\varphi_P}(\gamma) = \frac{1}{\gamma} - \frac{3}{2\gamma^2}(1 - P) \int_1^\infty \left(\frac{1}{a^3} - \frac{1}{a^5} \right) \frac{1 - e^{-\gamma a}}{1 - Pe^{-\gamma a}} \, da \qquad (3.26)$$

The variable $a = 1/\cos\theta$.

The last expression becomes equation 3.18 when $P = 0$. For $P = 1$ we obtain $\varphi_P(\gamma) = \gamma$ and $\sigma_0/\sigma_f = 1$, so that there is no size effect in the conductivity which is independent of the thickness of the film.

In the extreme cases of thick films ($\gamma > 1$) and thin films ($\gamma < 1$) equations 3.25 and 3.26 give the following relations [3.1, 3.2]:

$$\frac{\sigma_0}{\sigma_f} = \frac{\rho_f}{\rho_0} \approx 1 + \frac{3}{8\gamma}(1 - P), \qquad \text{for } (\gamma > 1) \qquad (3.27)$$

and

$$\frac{\sigma_0}{\sigma_f} = \frac{\rho_f}{\rho_0} \approx \frac{4}{3}\frac{1-P}{1+P}\frac{1}{\gamma[\ln(1/\gamma)+0.42]} \approx \frac{4}{3}\frac{1}{\gamma(1+2P)}\frac{1}{\ln(1/\gamma)}, \quad (3.28)$$

for $(\gamma \ll 1, P < 1)$.

If the parameter $P = 0$, equations 3.27 and 3.28 correspond to equations 3.19 and 3.20.

Figure 3.2 shows the dependence of the reduced resistivity ρ_f/ρ_0 of a thin film on the reduced thickness $\gamma = h/l$ calculated using the exact formula 3.25 for values $P = 0$ (curves 1) and also using the approximate formulas 3.27 and 3.28 for $P = 0$ (curves 3 and 2 respectively).

One can see from Figure 3.2 that one can use the approximate equations 3.19 and 3.20, or 3.27 and 3.28, with reasonable accuracy.

Figure 3.3 shows the dependence of the reduced resistivity of a thin film on the reduced thickness $\gamma = h/l$ for various values of the mirror parameter, P. If $P = 1$ the reduced resistivity is $\rho_f/\rho_0 = 1$ for all γ.

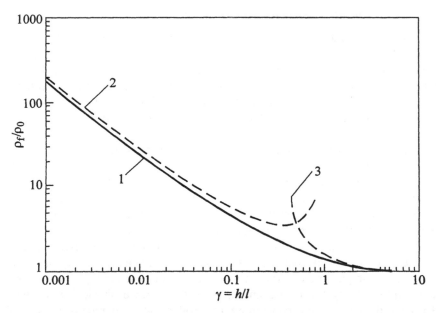

Figure 3.2 Dependence of the reduced resistivity of a thin film on the reduced thickness for the value of the mirror parameter $P = 0$ [3.3]. Curve 1 is calculated using the exact formula 3.25 while curves 2 and 3 are calculated using the approximate formulas 3.28 and 3.27 respectively.

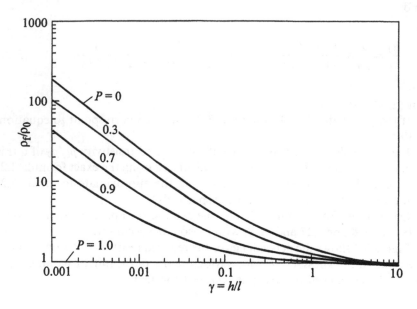

Figure 3.3 Dependence of the reduced resistivity of a thin film on the reduced thickness $\gamma = h/l$ for various values of the mirror parameter P.

3.1.3 Experimental results on the size effect

First we will consider the requirements for the samples and then we will outline the results of some experimental research on the size effect in metal films.

For good experimental results the thin film samples should be homogeneous, pure with a low concentration of structural imperfections and have plane and parallel surfaces. The thickness of the films should be comparable with, or smaller than, the length of the electron free path at the temperature of experiment.

Measurements of the size effect of the resistance are usually carried out on a set of films with increasing thickness and for which the electron free path and the concentration of imperfections and structural defects are identical. However this condition is difficult to fulfil since the microstructure may change as the film thickness increases such that it produces a change in the resistivity with thickness that is not connected with the size effect.

Equations 3.27 and 3.28 for the size effect in the film conductivity contain three experimental variables: ρ_0, l, P. Since each film can have very different values of these parameters, their correct determination for a particular film is possible only by the measurement of three or more physical quantities. However it is very time consuming to produce well characterised sets of samples. Therefore the basic parameters for the sequence of films of various thicknesses are assumed to be constant and then the experimental dependencies are compared with the theoretical curves constructed for various values of these parameters. The best values of the parameters ρ_0, l, P are determined from the best coincidence of the experimental and theoretical results [3.2, 3.3].

This means that the bulk conductivity of the film material, which is due to the scattering of the carriers by phonons and crystal lattice defects including impurities, is usually less than the real bulk metal conductivity because of the higher concentration in the films of crystal lattice imperfections and impurities.

It is best to carry out size effect experiments on films of the precious metals; silver and gold. Such films are most resistant to the effects of residual gases; they contain few impurities and do not oxidise much. Fuchs' theory is directly applicable for these metals since the Fermi surface is almost spherical and the mean free path is approximately isotropic along all directions. It is assumed that one electron per atom participates in the conduction and that its effective mass is equal to the mass of a free electron.

The measurement of the conductivity is best carried out at low temperatures ($T = 4.2\,\mathrm{K}$) since the electron mean free path becomes much larger than its value at room temperature and this allows experiments on thicker samples to be carried out. For these samples there is less change in the structure with changes in the thickness so that it is possible to obtain better results.

For comparison of the theory with the experimental data it is convenient to arrange equations 3.27 and 3.28 as [3.2, 3.3]:

$$\frac{1}{\rho_f h} \approx \frac{3}{4}\frac{1}{\rho_0 l}(1 + 2P)(\ln l - \ln h) \quad (\gamma \ll 1,\ P \ll 1); \tag{3.29}$$

$$\rho_f h \approx \rho_0[h + \frac{3}{8}l(1 - P)] \quad (\gamma > 1) \tag{3.30}$$

From equation 3.29 (the limit $\gamma \ll 1$) the graph of the dependence of $\rho_f(h)$ in the coordinates $(1/\rho_f h, \ln h)$ is given by a straight line with slope $-3/4(1/\rho_0 l)(1 + 2P)$, which intersects the x-axis at the point $\ln h_0 = \ln l$. It is thus possible to determine the values of l and $(1 + 2P)/\rho_0$ from this graph.

From equation 3.30 (the limit $\gamma \gg 1$) it follows that the graph of the dependence $\rho_f(h)$ in the coordinates $(\rho_f h, h)$ is given by a straight line with slope ρ_0 which crosses the y-axis at $(3/8)\rho_0 l(1 - P)$.

Thus, from the experimental graphs it is possible to determine the parameters: ρ_0, l, P. Since the parameter P is not very sensitive to the range of values $\gamma \gg 1$, no conclusions can be made about the magnitude of P from the data of graph (3.30).

Figure 3.4 shows the experimental results of Kadereit for thin gold films at a temperature of 4.2 K [3.3, 3.4]. The films were deposited on glass with a thickness in the range 60–1000 nm at different pressures of oxygen in the vacuum chamber. It can be seen from Figure 3.4 that an increase in the oxygen pressure during the film condensation results in a relatively small increase in the resistivity of the film. These results support the size effect theory. The continuous straight lines were obtained from the Fuchs' theory from formula 3.28 for $\gamma \ll 1$ and $P = 0$. The straight line (Au1) corresponds to

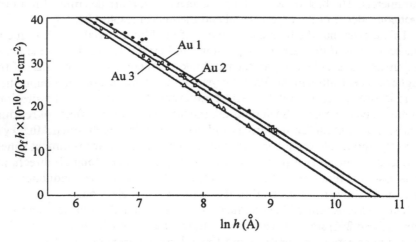

Figure 3.4 Dependence of the magnitude of $1/\rho_f h$ for gold films of thickness h at $T = 4.2\,\mathrm{K}$ [3.4]: The oxygen pressure in the vacuum chamber during film deposition had the following magnitudes: for Au1 – 10^{-6} Torr, for Au2 – 10^{-4} Torr, for Au3 – 10^{-3} Torr.

the value of the electron mean free path $l = 2.86\,\mu\mathrm{m}$ and the bulk resistivity of the films $\rho_0 = 0.0286\,\mu\Omega\,\mathrm{cm}$. These values assume one free electron per atom.

Experimental results for Au films at room temperature are given in Figure 3.5 [3.5]. The continuous straight line corresponds to the Fuchs theory for $\gamma > 1$ and $P = 0$ (equation 3.30) and gives a value for the electron mean free path $l = 42\,\mathrm{nm}$ that coincides with the length of the free path due to phonons. For these films the bulk resistivity $\rho_0 = 2.25\,\mu\Omega\,\mathrm{cm}$. These experimental results agree well with the Fuchs theory at $P = 0$ and indicate that there is completely diffuse scattering of the electrons at the surface of these polycrystalline gold films at $T = 4.2\,\mathrm{K}$ and $295\,\mathrm{K}$.

Let us now give some experimental data for the mirror parameter P [3.5]. The effect of surface scattering depends on the nature of the scattering mechanism in the film. Thus two extreme cases are possible: specular (elastic) reflection and diffuse reflection.

For specular reflection of the current carrier only the component of velocity that is normal to the surface of a film changes sign and the energy remains constant. Since in this case there is no loss of energy, or momentum in the conduction direction, the surface scattering does not affect the film conductivity. The surface is an ideal reflector.

For diffuse reflection the carriers go away from the surface with velocities which do not depend on the initial velocities. This change in the carrier momentum changes the film conductivity.

Specular reflection is expected for an ideal surface. The real surface of a film is determined by some irregularity that alters the properties of some of

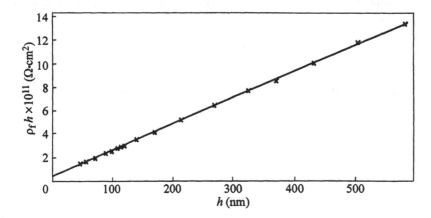

Figure 3.5 Size dependence of the magnitude of $\rho_f h$ on the thickness h for gold films at $T = 295\,\text{K}$: $\gamma \gg 1$, $P = 0$ [3.4].

the scattered carriers by an amount which depends on the type of the surface imperfections, their density and the effective cross-section for the scattering. Therefore for a real surface the scattering can be partially specular and partially diffuse so that the electrical properties of the film appear intermediate between the two extreme cases.

A value of $P = 0$ was obtained for films of the alkaline metals and polycrystalline aluminium, while for polycrystalline copper films $P = 0.47$ and for polycrystalline gold films $P = 0.2$. For single crystal gold films $P = 0.8$–0.9, which demonstrates high elastic reflection [3.5].

The mirror parameter can change with a modification of the surface electronic structure. Thus for very pure films it can decrease with even very small gas adsorption, which creates scattering centres on the surface.

3.1.4 The temperature coefficient of resistance of metal films

It was noted in Section 1.1.5 that the temperature coefficient of resistivity of thin continuous films is less than that of bulk metals because of the higher concentration of defects and impurities in films than in bulk conductors and also because of the influence of the size effect. This is because only the part of the resistivity in equations 1.4 and 1.5 caused by lattice scattering depends on the temperature. The temperature coefficient of resistivity of a film is given by $(1/\rho_f)(d\rho_b/dT)$ and is less than the temperature coefficient of the bulk metal $(1/\rho_0)(d\rho_b/dT)$.

Let us now consider the influence of the size effect on the temperature coefficient of resistivity of metal films. In equation 1.3 for the conductivity and resistivity, only the length of the free path depends on the temperature. Therefore the temperature dependence of the conductivity of the isotropic bulk metal is determined by the temperature dependence of $l(T)$, so that [3.2]:

$$\alpha_0 = \frac{1}{\rho_0}\frac{d\rho_0}{dT} = -\frac{1}{\sigma_0}\frac{d\sigma_0}{dT} = -\frac{1}{l}\frac{dl}{dT} \tag{3.31}$$

The temperature coefficient of the film resistance is determined by equation 1.9 so that:

$$\alpha_f = \frac{1}{\rho_f}\frac{d\rho_f}{dT} = -\frac{1}{\sigma_f}\frac{d\sigma_f}{dT} \tag{3.32}$$

For the ratio α_f/α_0 we obtain:

$$\frac{\alpha_f}{\alpha_0} = \frac{l}{\sigma_f}\frac{d\sigma_f}{dl} \tag{3.33}$$

Generally the theory of the size effect expressed by equation 3.25 gives:

$$\sigma_f = \sigma_0 \frac{\gamma}{\varphi_P(\gamma)} \tag{3.34}$$

for the film conductivity and we obtain the equality:

$$\frac{d\sigma_f}{dl} = \frac{d\sigma_0}{dl}\frac{\gamma}{\varphi_P(\gamma)} - \sigma_0\frac{d[\gamma/\varphi_P(\gamma)]}{d\gamma}\frac{h}{l^2} \tag{3.35}$$

After substitution of equations 3.34 and 3.35 into 3.33 we obtain the equation for the reduced temperature coefficient of the film resistance as [3.2]:

$$\frac{\alpha_f}{\alpha_0} = 1 - \varphi_P(\gamma)\frac{d[\gamma/\varphi_P(\gamma)]}{d\gamma} \tag{3.36}$$

Let us now derive expressions for the temperature coefficient of resistance in the cases of very thick and thin films. Initially we will consider the case of thick films ($\gamma > 1$). From equation 3.27 it is easy to obtain:

$$\sigma_f \approx \sigma_0\left[1 - \frac{3}{8\gamma}(1 - P)\right], \tag{3.37}$$

$$\frac{d\sigma_f}{dl} \approx \frac{d\sigma_0}{dl}\left[1 - \frac{3}{8\gamma}(1 - P)\right] - \sigma_0\frac{3}{8h}(1 - P) \tag{3.38}$$

After substituting the values of σ_f and $d\sigma_f/dl$ in to equation 3.33 and according to equation 3.31 $(l/\sigma_0)(d\sigma_0/dl) = 1$, we obtain:

$$\frac{\alpha_f}{\alpha_0} \approx 1 - \frac{3}{8}\frac{1 - P}{\gamma} \tag{3.39}$$

Thus at very large thickness ($\gamma \rightarrow \infty$) the temperature coefficient of resistance of a film tends to the value for the bulk material. This is independent of the value of P.

For the conductivity of very thin films ($\gamma \ll 1$), we have, from equation 3.28 the following:

$$\sigma_f \approx \frac{3}{4}\sigma_0\gamma\frac{1+P}{1-P}\ln\frac{1}{\gamma} \tag{3.40}$$

$$\frac{d\sigma_f}{dl} = \frac{3}{4}\frac{1+P}{1-P}\left[\frac{d\sigma_0}{dl}\gamma\ln\frac{1}{\gamma} + \sigma_0\frac{h}{l^2}\left(1-\ln\frac{1}{\gamma}\right)\right] \tag{3.41}$$

After substitution of σ_f and $d\sigma_f/dl$ into equation 3.33 we obtain an equation for the temperature coefficient of resistance for the case $\gamma \ll 1$ and $P \ll 1$:

$$\frac{\alpha_f}{\alpha_0} = \frac{1}{\ln(1/\gamma)} \tag{3.42}$$

Equation 3.42 shows that with a decrease of the film thickness the temperature coefficient of resistance also decreases. Note that the ratio α_f/α_0 is independent of the parameter P only for small values ($P \ll 1$).

The theoretical dependence of the ratio α_f/α_0 on the reduced thickness γ calculated using equation 3.36 is presented in Figure 3.6 (continuous curves) [3.2] for values of the parameter P. The experimental results for polycrystalline gold films (light circles) agree well with theory at $P=0$. The comparison of the experimental data with theory for the epitaxial gold films (dark circles) indicates that they show specular scattering. The best fit of the experimental results to theory for these films is for $P=0.5$ at $\gamma > 0.6$ (see Figure 3.6). Note that formulas 3.36 and 3.42 can only be used while the film is continuous.

In films of the refractory metals (tantalum, titanium, molybdenum and chromium) a negative temperature coefficient of resistivity is sometimes observed even for a film thickness much greater than the critical one so that they are continuous. These metals have a high affinity with oxygen. Therefore at a low condensation rate or if the films condense in a poor vacuum these metals can interact with oxygen, or with nitrogen in the case of tantalum. Then thin dielectric layers of oxides or nitrides can form on the boundaries of the grains. Some grains in such films will become isolated from the others by the dielectric interlayer with a thickness about 0.1–1 nm. Such a film is a heterogeneous system in which the mechanism of conduction is not metallic but is a thermally activated mechanism which is similar to the conduction mechanism of island films considered in Chapter 2. In the island films the grains are isolated by vacuum spaces while in these films they are isolated by dielectric layers. The temperature coefficient of film resistance for such films is negative and is determined by an equation such as 2.21. However the resistivity of such films will be considerably smaller than that of island films since the resistance is determined by the contribution of the metallic conductivity.

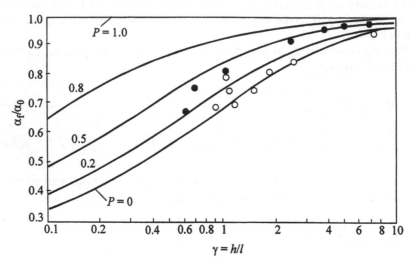

Figure 3.6 The theoretical dependence of the normalised temperature coefficient of resistivity of gold films on the normalised thickness for various values of the mirror parameter P (continuous curves). The experimental results are represented by dots. The light circles are for polycrystalline films. The dark circles are for epitaxial films [3.2].

3.1.5   *The temperature coefficient of resistance of semi-metal films*

Semi-metals are substances with electrical properties intermediate between metals and semiconductors. Bismuth, antimony, arsenic and graphite are considered to be semi-metals. They have a weak overlap of the valence and conduction bands. For example, for bulk bismuth the overlap of the bands is about 0.02 eV. Thus semi-metals remain conductors down to absolute zero, although they have only a small concentration of current carriers compared with metals. This equals 10^{17}–10^{20} cm^{-3} at room temperature [3.6]. For bismuth the concentration of both electrons and holes is $n = p \approx 2.75 \times 10^{17}$ cm^{-3}. The current carriers in semi-metals have a high mobility and a small effective mass. For example, for bismuth the electron mean free path at room temperature is equal to $l \approx 2\,\mu$m. Because of this the semi-metals are very suitable for research into size effects.

In bulk semi-metals the conductivity is equal to $\sigma = e n \mu_{n} + e p \mu_{p}$ [3.6] where μ_{n} and μ_{p} are the mobilities of the electrons and holes respectively ($\mu \propto l$). For electron conduction in a semi-metal equation 1.3 holds. With increasing temperature the number of carriers increases and the mean free path, l, decreases. But l decreases faster than n increases so that, according to equation 1.3, the resistivity will increase with temperature monotonically to give a positive temperature coefficient of resistivity ($\alpha_0 > 0$). For bismuth $\alpha_0 \approx 5 \times 10^{-3}\,\mathrm{K}^{-1}$.

However, in bismuth films with a continuous structure a negative temperature coefficient of resistivity is observed in a defined temperature interval. The experimental temperature dependence of the resistivity of thin Bi films of different thickness are given in Figure 3.7 [3.7]. The films were deposited on glass substrates. The thickness of the films in these experiments was varied from 2.4 to 42.0 μm. It can be seen that the dependence of the resistivity at high temperatures ($T = 300$–400 K) for films of different thickness do not differ from each other within the accuracy of measurement of the thickness. The temperature coefficient of resistance in this temperature range is close to that of bulk material ($\alpha_f > 0$). At lower temperatures ($T \approx 100$–200 K) an increase in the resistivity with a decrease in the temperature is observed. A negative temperature coefficient occurs over a wider temperature range for thinner films.

It is possible to explain this result qualitatively with the help of Fuchs' theory. In films where $h \ll l$, the mean free path decreases much slower than the concentration of carriers, n, increases as the temperature increases. In this case the resistivity will decrease as the temperature increases, to give a negative temperature coefficient of resistivity for semi-metal films over some temperature interval. At high enough temperature the mean free path

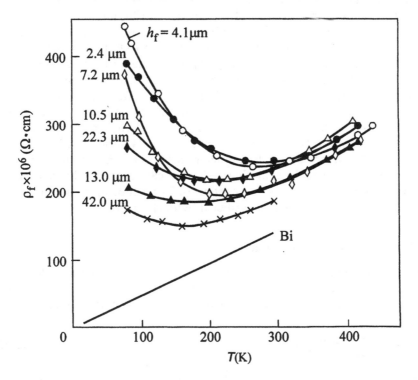

Figure 3.7 The temperature dependence of the resistivity of thin bismuth films of different thickness [3.7]. The bottom curve is for bulk bismuth.

decreases so that $h > l$ and the temperature coefficient of the semi-metal film becomes positive, as for bulk bismuth.

3.1.6   Full description of Fuchs' theory

As mentioned earlier, formulas 3.27 and 3.28 were obtained using a simplified model. In Fuchs' theory the mirror parameter is a constant which does not depend on the de Broglie wavelength of the electron or on the angle of impact of an incoming electron on the surface of the film. The mirror parameter can depend on the angle of impact because of the anisotropy of the dispersion. Also the mirror parameter may be different for the two surfaces of the film [3.2].

In thin films with diffuse reflection from the film surface, the current is mainly determined by the electrons moving almost parallel to the surfaces. Measurements of the temperature dependence of the resistance of thin wires showed [3.8] that the temperature dependent part, ρ_b, connected with lattice scattering is a function of the size of the sample. This was explained by the influence of the surface on the frequency of the electron–phonon collisions. The effective frequency of the electron–phonon collisions in a thin wire or film can appear to be greater than in the bulk metal. In the bulk metal the small angle scattering gives only a small contribution to the resistivity. However, in films the small angle phonon scattering can lead to additional collisions of the electrons with the surface for those electrons moving almost parallel to the surface. This will magnify the resistance in the case of diffuse scattering. Thus the carrier mean free path with electron–phonon dispersion will decrease with a decrease in the film thickness. The effectiveness of small angle electron scattering at low temperatures, $T/\theta_D \leq h/l$ (θ_D is the Debye temperature), results in this component of resistivity varying as T^3 [3.5] while, according to Bloch's law, the bulk resistivity varies as T^5 [3.6].

Thus at low temperatures deviations from Matthiessen's rule (equation 1.5) are observed so that the total film resistance cannot strictly be divided into phonon and impurity, $\rho_d + \rho_s$, resistances. The temperature dependence of resistance will differ from that predicted by Fuchs' theory.

3.2   Hall effect in thin films

When a conductor is placed in a magnetic field that is perpendicular to the direction of the current, an electrical field, E_y, directed perpendicular to both the current and magnetic field, appears across the sample (Figure 3.8):

$$E_y = \frac{j_x B_z}{ne} = R_H j_x B_z \tag{3.43}$$

where j_x is the current density through the conductor in the x-direction; B_z is the magnetic induction; n is the carrier concentration; R_H is the

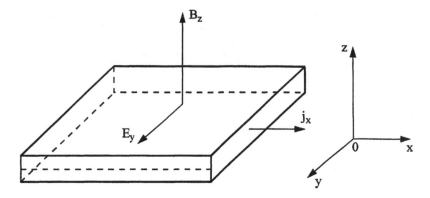

Figure 3.8 The Hall effect in a thin film.

Hall constant or coefficient. This coefficient is therefore given by the following equation:

$$R_H = \frac{1}{ne} \qquad (3.44)$$

The Hall coefficient depends on the carrier concentration and is therefore a good way to measure this. Since the sign of the transverse potential difference depends on the sign of the charge carriers, it is also possible to determine the type of carrier in a metal. This is especially interesting in the study of the properties of the transition metals. For example, it has been established that in films of yttrium and scandium with a thickness of $h > 10$ nm it is possible to consider only the electron component of conduction since the contribution of the hole component is negligible.

In thin films the scattering of electrons at the surface of the film influences the Hall effect. The theory of the size effect gives the following equations for the Hall coefficient in thin films, R_{Hf}, for the two extreme cases [3.2]:

$$R_{Hf} \approx R_{H0} \quad (\gamma > 1), \qquad (3.45)$$

$$R_{Hf} \approx R_{H0} \frac{4}{3} \frac{1-P}{1+P} \frac{1}{\gamma [\ln(1/\gamma)]^2} \quad (\gamma \ll 1, P \ll 1) \qquad (3.46)$$

where R_{H0} is the Hall constant for the bulk material.

The extreme case of $\gamma > 1$ is obeyed down to values of γ close to unity. The theoretical dependence of the Hall coefficient, normalised to the bulk Hall coefficient, with the normalised thickness, γ, for various values of the mirror parameter, P, is shown in Figure 3.9 (left axis). This increase in the Hall coefficient for thin films has been seen experimentally [3.2] so that as

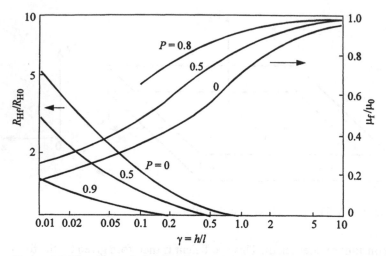

Figure 3.9 The theoretical dependence of the Hall coefficient, R_{Hf}, normalised to the bulk Hall coefficient, R_{H0} (left axis) and the Hall mobility of carriers, μ_f, normalised to the bulk magnitude of Hall mobility μ_0 (right axis) for a thin film with reduced thickness, γ, for various values of the mirror parameter, P, in a weak magnetic field [3.2].

the film thickness decreases an apparent decrease in the apparent concentration of the carriers is observed.

In the free electron model, the conductivity of a sample of any thickness is given by the formula:

$$\sigma = ne\mu \tag{3.47}$$

where μ is mobility of the carriers. Then the Hall coefficient according to the equations 1.1, 3.44 and 3.47 is given by:

$$R_H = \mu\rho \tag{3.48}$$

An equation for the Hall mobility of carriers in a film, μ_f, is obtained from equation 3.48 after substitution in it the value of R_{Hf} from formulas 3.45 and 3.46 and ρ_f from equations 3.27 and 3.28:

$$\mu_f \approx \frac{\mu_0}{1 + \dfrac{3}{8\gamma}(1-P)} \quad (\gamma > 1), \tag{3.49}$$

and

$$\mu_f \approx \mu_0 \frac{1}{\ln(1/\gamma)} \quad (\gamma \ll 1) \tag{3.50}$$

where $\mu_0 = R_H\rho_0$ is the bulk magnitude of the Hall mobility.

The theoretical curves for the variation of μ_f/μ_0 with γ for various values of P are given in Figure 3.9 (right axis). It can be seen that the size effect modifies the Hall mobility for $\gamma \geq 1$ more strongly than it does the Hall coefficient. Thus, if the Hall coefficient in thin films is measured as well as the conductivity, it is possible to find out how the concentration of the current carriers varies and also to obtain information about the sign of the carriers and their scattering properties at a surface.

3.3 Quantum size effect in electrical conduction

The quantum size effect is observed when the de Broglie wavelength of an electron is comparable with the film thickness [3.3, 3.5]. Then the transverse motion of the particles is quantised so that the projection of the electron quasi-momentum, $\bar{P}_r$, on the normal to the plane of the film can only accept discrete values. For quasi-particles with quadratic dispersion in a rectangular potential well with a flat bottom and infinitely high walls, the projection of the quasi-momentum on the z-axis, normal to the plane of a film, is [3.5]:

$$|P_z| = \frac{\pi\hbar}{h}n, \quad n = 1, 2, 3, \ldots \tag{3.51}$$

where $\hbar = h/2\pi$, and h is the Plank constant.

In these calculations specular reflection of electrons from the surface is usually assumed ($P = 1$). Partially diffuse reflection produces a decrease in the observable effect [3.3].

The thickness of the film at which quantum effects will be observed, h', is the order of magnitude of the de Broglie wavelength, λ_D, of the electron:

$$h' \sim \lambda_D = \frac{2\pi\hbar}{\sqrt{2mE_e}} \tag{3.52}$$

where E_e is the energy of the electron.

The film thickness at which quantum effects will be observed for metals is $h' \approx 10^{-8}$ cm, and for degenerate semiconductors or semi-metals, $h' \approx 10^{-5}$ cm. This means that the quantum size effect cannot be observed in metals since at a film thickness of $h' \approx 10^{-8}$ cm the film is a single atomic layer. Therefore the quantum size effect is found only in films of degenerate semiconductors or semi-metals.

Spatial quantisation produces a splitting of the energy levels into two subbands. For the simple model of a film, without including any interactions between the conduction electrons in the volume and using an isotropic effective mass the spectrum of the electron energy is given by [3.5]:

$$E_e = \frac{P_{//}^2}{2m} + \frac{\pi^2\hbar^2}{2mh}n^2 \tag{3.53}$$

Here a component of the quasi-momentum $P_{//}^2 = \hbar^2(k_x^2 + k_y^2)$ where k_x and k_y are the values of the wave vector of an electron in the x- and y-directions.

Thus the power spectrum of a quasi-particle in a film equation 3.53 is quasi-discrete and is broken into sub-bands. The distance in energy between the states described by the quantum numbers n and $n+1$ in the adjacent sub-bands is:

$$\Delta E_{en} = \frac{\pi^2 \hbar^2}{2mh^2}(2n+1) \tag{3.54}$$

The minimum energy, E_{e1}, is determined from equation 3.53 for $P_{//} = 0, n = 1$:

$$E_{en} = \frac{\pi^2 \hbar^2}{2mh^2} \tag{3.55}$$

For semi-metal films under the conditions for the quantum size effect the conduction band (CB) and valence band (VB) are split into discrete levels. Thus the lowest energy of the conduction band is moved above the bottom of the conduction band of a bulk sample by an amount:

$$\Delta E_{e1} = \frac{\pi^2 \hbar^2}{2mh^2} \tag{3.56}$$

and the highest energy of the valence band is moved down from the top of the valence band of the bulk sample by:

$$\Delta E_{h1} = \frac{\pi^2 \hbar^2}{2m_h h^2} \tag{3.57}$$

where m_h is the effective mass of a hole.

If the film thickness is small enough, the sum $\Delta E_{e1} + \Delta E_{h1}$, can become equal to the magnitude of the band overlap in a bulk semi-metal (ΔE) and the overlapping of the bands disappears. At still smaller thickness a forbidden band appears and semi-metal becomes a semiconductor.

The thickness at which the band overlap disappear is equal to [3.3]:

$$h' = \pi\hbar/\sqrt{2M\Delta E} \tag{3.58}$$

where $M^{-1} = m^{-1} + m_h^{-1}$.

For $m_h \gg m_e$ we obtain:

$$h' = \frac{\pi\hbar}{P_{//}} = \frac{\lambda_D}{2} \tag{3.59}$$

i.e. the thickness h' is equal to half the de Broglie wavelength of an electron.

In Figure 3.10 the dependence of the electron energy on the projection of the wave vector on the x-axis, k, at fixed projections of k_y and k_z, are given for a semi-metal film under the conditions for the quantum size effect. Electrons in the conduction band and holes in the valence band fill the shaded areas. Figure 3.10a corresponds to the bulk semi-metal. The bands overlap by an energy interval ΔE which is the magnitude of the band overlap for bulk semi-metals. Figure 3.10b corresponds to a thin film of thickness $h > h'$. Here the overlap of the bands is smaller and the conduction and valence bands are split into discrete sub-bands (not shown on the figure). Figure 3.10c corresponds to a thin film of thickness $h < h'$, when the band overlap disappears, an energy gap appears and the semi-metal becomes a semiconductor.

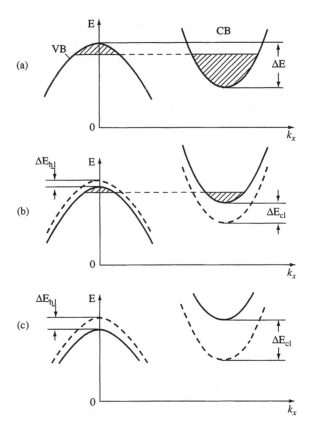

Figure 3.10 The energy bands for a semi-metal under the conditions for the quantum size effect: (a) bulk semi-metal; (b) a film with a thickness $h > h'$; (c) a film with a thickness $h < h'$. The dotted lines are the dispersion curves for a bulk semi-metal [3.3].

Sandomirskiy obtained a formula for the resistivity of a film for a film thickness larger than the thickness h' and for very low temperatures $(kT/\Delta E \ll 1)$ [3.9], the results are shown in Figure 3.11 [3.3]. The resistivity of the film shows oscillations with thickness with a period h'. For bismuth $h' = 39$ nm.

Experimental observation of quantum size effects were first reported by Ogrin, Lutsiy and Elinson [3.10]. They investigated the size dependence of the resistivity and the Hall coefficient for bismuth films at temperatures 4.2 K, 78 K, and 300 K. Bismuth was deposited on a heated mica substrate. The resistivity, Hall coefficient and the Hall mobility showed oscillations with a period of about 40 nm at temperatures 4.2 K and 78 K. This is in good agreement with the value $h' = 39$ nm predicted by theory.

The experimental dependence of a resistivity of bismuth films on thickness at temperatures of 4.2 K, 78 K and 300 K are given in Figure 3.12. The amplitude of the oscillations decrease with increasing temperature and film thickness. At room temperature the relation $(kT/\Delta E \ll 1)$ does not hold and the oscillations are not seen (for bismuth $\Delta E = 0.02$ eV).

For a thickness of less than 40 nm the resistivity increases rapidly with a decrease in thickness. This can be due to the change of the semi-metal into a semiconductor or because the film becomes discontinuous (see Section 1.5.4).

The study of the quantum size effect in semi-metals and degenerate semiconductors is of interest since it is one of only a few examples of observations of quantum mechanical effects in macroscopic samples.

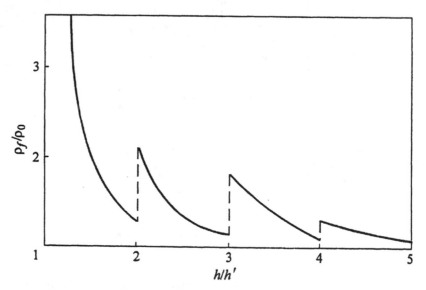

Figure 3.11 Calculated dependence of the reduced resistivity on the reduced thickness for a bismuth film [3.3, 3.9].

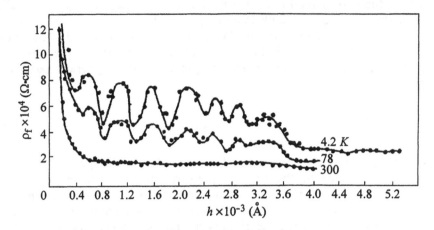

Figure 3.12 The dependence of the resistivity of bismuth films on thickness at different temperatures [3.3, 3.11].

3.4   Size effects in the structure of thin films

3.4.1   The phase size effect

Studies of the structure of metal films by electron and X-ray diffraction have show that non-equilibrium structural phases are frequently found in thin films which are absent in bulk metals. In films of Ta, Nb, W and Mo the anomalous face-centred cubic (f.c.c.) crystallographic structure has been detected instead of the body-centred (b.c) equilibrium structure which is inherent in bulk metals [3.5].

In Ta films two structural modifications have been observed, which are known as α-Ta and β-Ta. The α-Ta films have a body centred structure and a low resistivity $\rho_f \approx 20\,\mu\Omega\,cm$. The temperature coefficient of resistivity of these films is $\alpha_f \approx 2 \times 10^{-3}\,K^{-1}$. The β-Ta films have a tetragonal structure, which is only observed in films [3.11]. The resistivity is $\rho_f \approx 200\,\mu\Omega\,cm$ and the temperature coefficient of resistivity of these films is close to zero. The tetragonal structural modification, β-Ta, is only observed in films [3.11].

For continuous metal films of Cu, Ag, Au, Ta the crystal lattice constant was found to be larger than in bulk crystals [3.2]. Further studies showed that the increase in the period of the crystal lattice in all cases was the result of the influence of impurity atoms, especially gas impurities.

A decrease in the crystal lattice period from the value for the bulk metal has been observed. It has been established experimentally that in island films the lattice constant could either increase or decrease as the island size decreases but when the influence of impurities of residual gases was reduced to a minimum, the period of the lattice was always smaller than the value for the bulk material [3.4]. With a decrease of the crystal size a greater decrease in the lattice period was observed. This effect is connected with the influence of the Laplace surface pressure. For a small particle the Laplace surface pressure for a spherical particle is given by:

$$P_1 = 2\sigma_S/r \tag{3.60}$$

where σ_S is the surface tension and r is the radius of the sphere.

The magnitude of the Laplace pressure for a particle with a radius $r = 2\,\text{nm}$ using a typical value for metals of $\sigma_S \approx 1\,\text{J m}^{-2}$ is $P_1 \approx 10^9\,\text{Pa} \approx 10^4\,\text{Atm}$. Under the influence of such a high surface pressure the volume of a particle appears smaller, and that reduces the period of the lattice. Thus in small particles of high-pressure, high density phases are possible although they are absent in bulk materials. This effect is shown mainly in island films containing particles of very small size. This phenomenon is called the phase size effect.

The origin of non-equilibrium phases in thin films is the small size of the particles created during the initial stage of film formation. The further growth of the particles during the film condensation should be accompanied by a phase change from the non-equilibrium to the equilibrium phase. However under real conditions the transformation is probably delayed since the phase which is formed first will be reproduced by epitaxy during further condensation of the film. Therefore the crystal structures of the initial stages of the condensation process can appear in continuous films as well as in island films.

The relative decrease in the volume of small particles and consequently of the crystal lattice period under the influence of the surface tension is determined by the effect of the Laplace pressure (equation 3.60) on the coefficient of compressibility, ξ [3.12]:

$$\frac{V_0 - V}{V_0} \approx \frac{3(a_0 - a)}{a_0} = \xi P_1 = \frac{2\xi\sigma_S}{r} \tag{3.61}$$

where V_0 and a_0 are the volume of a particle and the lattice constant for the case when the influence of the surface energy can be neglected, and V and a are the volume and lattice constant in the calculation of the Laplace pressure.

For metals the compressibility modulus $\xi \approx 10^{-11}\,\text{m}^3\,\text{J}^{-1}$. For a spherical particle with radius $r = 2\,\text{nm}$ the relative modification of the lattice constant according to equation (3.61) gives $(a_0 - a)/a_0 \approx 0.33\%$. Such a magnitude can easily be detected using diffraction methods.

3.4.2 The effect of the thickness of continuous films on the melting temperature

The surface energy causes the phase transition temperature to depend on the film thickness and on the size of small crystals. Let us consider this effect using the example of the dependence of the melting temperature of continuous films on the thickness. Calculations and experiments show that this can be significant. In very thin films ($h \leq 10$ nm) the decrease can be by tens and even hundreds of degrees. In films of Pb, Sn, Bi and Ag with a thickness of $h \approx 5$ nm the decrease in the melting temperature was observed to be 41, 30, 23 and 200 °C respectively [3.5]. It is easy to demonstrate this effect.

Let a solid phase, 1, exist at low temperatures and a liquid phase, 2, exist at high temperatures. Their thermodynamic potentials are equal at the equilibrium point of the solid and liquid phases. A condition for the equality of the isobaric–isothermal potentials, Gibbs energy, of the two phases of a film, or small particles, allowing for the surface energy is written as:

$$H_1 - T_1 S_1 + \sigma_{S1} \frac{A_S}{V_f} = H_2 - T_1 S_2 + \sigma_{S2} \frac{A_S}{V_f} \qquad (3.62)$$

Here H_1, H_2 are the enthalpies and S_1, S_2 are the entropies per unit volume, T_1 is the melting temperature of the film, σ_{S1}, σ_{S2} are the surface tensions for the solid and liquid phases respectively, A_S is the surface area and V_f is the volume.

For bulk crystals the surface energy can be neglected. Then at the equilibrium point of the solid and liquid phases we have:

$$H_1 - T_0 S_1 = H_2 - T_0 S_2 \qquad (3.63)$$

where T_0 is the melting temperature for the bulk crystal. From equation 3.63 let us calculate the magnitude:

$$S_2 - S_1 = \frac{H_2 - H_1}{T_0} = \frac{Q}{T_0} \qquad (3.64)$$

where Q is the heat of the phase transition from solid to liquid. By substituting equation 3.64 into equation 3.62 we obtain the equation:

$$1 - \frac{T_1}{T_0} = \frac{(\sigma_{S1} - \sigma_{S2}) A_S}{Q V_f} \qquad (3.65)$$

and for the difference $\Delta T = T_0 - T_1$ between the melting temperatures of the bulk crystal and the film we have:

$$\Delta T = \frac{(\sigma_{S1} - \sigma_{S2}) A_S T_0}{Q V_f} \qquad (3.66)$$

For the film $A_S/V_f \approx 2/h$ so that we obtain:

$$\Delta T = T_0 - T_1 = 2\frac{(\sigma_{S1} - \sigma_{S2})T_0}{Qh} \tag{3.67}$$

Because the surface of a film is not smooth the magnitude A_S/V_f can be more than $2/h$.

For a particle in the form of a sphere $A_S/V_f = 3/r$ where r is the radius of the sphere, so that:

$$\Delta T = T_0 - T_1 = 3\frac{(\sigma_{S1} - \sigma_{S2})T_0}{Qr} \tag{3.68}$$

The modification of the phase transition temperature can be quite significant. For the melting of metals, taking typical values $\sigma_{S1} - \sigma_{S2} \approx 10^{-1}\,\mathrm{J\,m^{-2}}$, $Q \approx 3 \times 10^8\,\mathrm{J\,m^{-3}}$ and for a film of thickness $h = 4\,\mathrm{nm}$ with an ideally smooth surface from equation 3.67 it follows that:

$$\Delta T \approx \frac{T_0}{6} \tag{3.69}$$

For silver $T_0 = 1200\,\mathrm{K}$ and from equation 3.68 we obtain a decrease of the melting temperature for a film of thickness $4\,\mathrm{nm}$ of $\Delta T = 200\,\mathrm{K}$. For small crystals the decrease of the melting temperature will be even more noticeable. For spherical particles of radius $r = 2\,\mathrm{nm}$ we obtain $\Delta T = T_0/2$ so that the melting temperature is halved.

Thus we see that the melting temperature of films depends on the thickness, is below the melting temperature of bulk crystals and is controlled by the influence of the film surface energy. The melting temperatures of films deposited on substrates of various materials can differ because of the difference in the surface energies of the film–substrate interface and also because of the difference in the structure of the films. The conditions during the condensation of the film also affect the melting temperature.

The experimental melting temperatures of films of silver and copper condensed on to an amorphous carbon substrate at a temperature $T_c \approx 500\,\mathrm{K}$, prepared and annealed in a vacuum with a residual gas pressure of 10^{-8}–$10^{-7}\,\mathrm{Pa}$ are shown in Figure 3.13 [3.13]. The measurement of the melting temperature of the film was carried out with the help of an electron microscope.

If the film's condensation temperature is lower than $450\,\mathrm{K}$ or higher than $850\,\mathrm{K}$, the melting temperature of the films is noticeably different from the values given in Figure 3.13. This is explained by differences in the grain size and by the lack of perfect smoothness of the film surface. In this connection the experimentally found dependencies $T_1(h)$ for continuous films cannot be

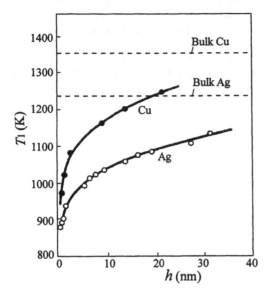

Figure 3.13 Experimental dependence on the thickness of the melting temperature T_1 of silver and copper films. The dotted lines show the
melting temperatures of bulk metals [3.13].

accounted for by the simple relation given in equation 3.67. The magnitude
of $\Delta \sigma_S = \sigma_{S1} - \sigma_{S2}$ is found to depend on the film thickness.

Literature for Chapter 3

3.1   Fuchs, K., "The conductivity of thin metallic films according to the electron
      theory of metal", *Proc. Camb. Phil. Soc.*, **34** (1938) 100–109.
3.2   Chopra, K., *Thin Film Phenomena*, J. Wiley, New York, 1969.
3.3   Larsen, D.C., "Size effects in conductivity of the thin metallic films and wires",
      Physics of Thin Films, Francombe, M.H. and Hoffman, R.W. (eds), Academic
      Press, New York and London, **6** (1971) 81–149.
3.4   Kadereit, H.G., *Thin Solid Films*, **1** (1967) 109.
3.5   Komnik, Yu.F., *Physics of Metallic Films: Size and Structural Effects* (in
      Russian), Atomizdat, Moscow, 1979.
3.6   Kittel, C., *Introduction to Solid State Physics*, J. Wiley, New York, 1996.
3.7   Kaiydanov, V.I. and Regel, A.R., "About influence of thickness of bismuth
      films on their electrical properties", *J. Tech. Phys.*, **28** (1958) 402–411 (in
      Russian).
3.8   Olsen, J.I., "Magnetoresistance and size effects in indium at low temperatures",
      Helv. Phys. Acta, **31** (1958) 713.
3.9   Sandomirskiy, V.B., "Quantum effect of sizes in film of semi-metals", *JETP*, **52**
      (1967) 158 (in Russian).

3.10 Ogrin, Yu.F., Lutskiy, V.N. and Elinson, M.I., "About observation of quantum size effects in Bi films", *JETP Lett.*, **3** (1966) 114 (in Russian).
3.11 *Handbook of Thin Film Technology* Maissel, L.I. and Glang, R. (eds), McGraw Hill, 1970.
3.12 Girifalko, L.A., *Statistical Physics of Materials*, J. Wiley, New York, 1973.
3.13 Gladkich, N., Niedermayer, R. and Spigel, K., "Nachweis großer Schmelzpunkterniedrigung bei dunnen Metallschichten", *Phys. Status Solidii*, **15** (1966) 181.

4 Internal mechanical stresses and voids in thin metal films

4.1 Biaxial stress in a thin film

Internal mechanical stresses (macro-stresses) appear in metal films after their condensation and during their processing and storage. These stresses are sustained by the interaction of the film with its substrate [4.1]. The magnitude of these stresses can reach the strength of the film. This can then result in the spontaneous destruction of the film by a crack or by its separation from the substrate. These mechanical stresses can also influence the electrical properties of the films. Therefore a study of mechanical stresses in thin films is of interest for the practical applications of films and especially for integrated circuits.

The internal stresses can be tensile (where the film tries to reduce its size in the directions parallel to the surface of the substrate) and are taken as positive in sign. In the opposite case the stresses are compressive and are taken as negative. It has been established that the strength (yield point) of thin films is hundreds of times larger than the strength of the bulk material. The high strength of thin films is explained by the high density of dislocations which are strongly fixed to point defects or to the surface of the film so that their movement under deformation is hampered. Macro-stresses can change if any heat treatment is given to the films, or during storage. They depend on the film's microstructure, impurity concentration and the imperfections of the crystalline lattice.

When studying the structural non-equilibrium of films one must consider any micro-stresses σ_m (or micro-strains ε_m) as well as the macro-stresses. A micro-deformation is the local deformation within a small volume of the crystal. The micro-stresses can be oriented, for example relative to the plane of the film, or randomly directed [4.2]. When considering macro-stresses it is assumed that they are in balance over the whole volume of the film, whereas when considering micro-stresses it is assumed that they are in balance within each grain or local volume. The magnitude of the micro-stresses at the boundaries of a grain can considerably exceed the level of the macro-stresses within a film [4.2].

The stress condition of a solid is described by a symmetrical orthogonal stress tensor of second rank. For the case of biaxial stress, when the stress lies in the plane of the film, the stress tensor reduces to a diagonal form, which looks like [4.3]:

$$T_\sigma = \begin{bmatrix} \sigma_1 & 0 & 0 \\ 0 & \sigma_2 & 0 \\ 0 & 0 & 0 \end{bmatrix} \tag{4.1}$$

where σ_1, σ_2 are the mechanical stresses in the two mutually orthogonal directions, the principal axes.

The biaxial stress condition is characteristic of a thin film which is subject mainly to forces imposed by the substrate. The force vector lies in the plane of the film so that there are no stresses perpendicular to the surfaces of the film. For a symmetrical biaxial stress in an isotropic film $\sigma_1 = \sigma_1 = \sigma$. In this case Hooke's law becomes:

$$\sigma = \frac{E_f}{1 - \nu_f} \varepsilon_f \tag{4.2}$$

where E_f is Young's modulus of the film material, ν_f is its Poisson ratio and ε_f is the relative deformation, or strain, of the film.

If the film is deposited on a thin substrate the substrate will curve because of the internal mechanical stresses in the film. If these stresses are tensile the substrate is curved so that the free surface of the film appears concave. The film compresses the substrate while the substrate expands the film. When there are compressive stresses in the film the free surface of the film is convex. This is shown in Figure 4.1.

If various physical and chemical changes occur in a film, the film volume is altered so that, if the film strongly fixed to its substrate, macro-stresses are generated. Therefore, when studying the mechanisms which generate internal macro-stresses in films, it is convenient to consider the various physical and chemical processes from the point of view of how they modify

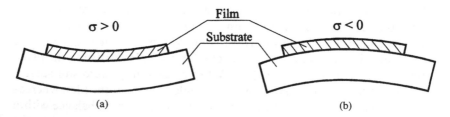

Figure 4.1 Curving of a substrate with a film on one side due to the internal stresses in the film: (a) for a tensile film stress; (b) for a compressive stress.

the density of the film [4.2]. Processes that produce a decrease in the density will decrease the tensile macro-stresses, or increase the compressive stresses, as long as the substrate acts against an increase in the film volume. For an increase in the density the tensile stresses increase, or the compressive stresses decrease in absolute value.

Before considering the physical mechanisms that produce internal mechanical stresses in films, we will give a classification of the defects found in solids.

4.2 Classification of defects in solids and thin films

Crystalline solids in thermodynamic equilibrium contain defects in their crystal lattice which considerably influence their physical properties. As well as equilibrium defects, there can also be non-equilibrium defects. The defects of a solid lattice can be classified by their geometry:

1 Point defects are vacancies and interstitial atoms;
2 Linear defects are dislocations and lines of introduced atoms and vacancies;
3 Area defects are the boundaries of grains and packing faults;
4 Volume defects are voids, cracks and scratches whose linear size exceeds the inter-atomic distance.

4.2.1 Point defects in solids and thin films

The essential difference between point defects and defects of other types is that they can exist in a crystal in thermodynamic equilibrium.

Away from equilibrium, the point defects of greatest interest are vacancies since their energy of formation, the removal of an atom from its lattice point to infinity, is much lower than the energy needed to move an atom into an interstitial position. Therefore the equilibrium density of vacancies in crystals is much larger than that of atoms in interstitial positions. When the concentration of vacancies in a solid is equal to the thermodynamic equilibrium concentration then the free energy of the crystal, as a function of the vacancy concentration, is at a minimum.

Most of the other defects are not in thermodynamic equilibrium so that their presence in a crystal increases the free energy of the crystal and their concentration often considerably exceeds the equilibrium value.

The density of equilibrium vacancies in a perfect crystal, N_v^0, at a temperature, T, is determined by the formula [4.4, 4.5]:

$$N_v^0 = N_a \exp(-u_v/kT) \tag{4.3}$$

where N_a is the density of lattice sites (atoms) in the crystal lattice and u_v is the energy of vacancy formation, which is equal to the work needed to remove an atom from the crystal. This work is determined by the number of chemical bonds that need to be broken in this process. The formation of a vacancy in an ideal crystal requires the rupture of about half of the bonds to an atom in a crystalline lattice, on the assumption that the atom that leaves the bulk goes to the surface of the crystal or a void. The energy of vacancy formation for different metals is $u_v = 0.7$–2.5 eV [4.4, 4.5]. This gives the energy due to one bond of an atom in a crystalline lattice as $u_{vl} = 0.1$–0.5 eV.

For example, for aluminium $u_v = 0.76$ eV[4.5], $N_a \approx 10^{22}$ cm^{-3}. Near the melting temperature (for aluminium $T_{mel} = 933$ K) the relative concentration of vacancies in a bulk material is, from equation 4.3:

$$n_v^0 = \frac{N_v^0}{N_a} = \exp(-u_v/kT) \tag{4.4}$$

equal to about 0.01 at.% ($N_v^0 \approx 10^{18}$ cm^{-3}). At room temperature, $T_0 = 300$ K, and $n_v^0 \approx 10^{-10}$ at.% ($N_v^0 \approx 10^{10}$ cm^{-3}).

Thin films, particularly those formed by evaporation-condensation in a vacuum, can be highly defected. Films crystallised from a strongly saturated vapour contain a high concentration of non-equilibrium defects. This concentration can exceed the concentration of defects in a bulk sample by several orders of magnitude.

The relative concentration of non-equilibrium vacancies, n_v, in a freshly deposited film can reach a value $n_v \approx 1$ at.% so that the density of non-equilibrium vacancies can reach a value of $N_v \approx 10^{-20}$ cm^{-3} [4.2]. Such a high concentration of vacancies can be explained by the capture of vacancies in the growing layer of the film. The conditions for condensation can be such that the vacancies that form have no time to diffuse out to the condensation front and hence leave the growing solid. Thus, an extremely large number of vacancies can remain in a growing film and be included in its volume. Thus there are many non-equilibrium vacancies that are available to leave the film by movement to various sinks during the storage or annealing of the film.

4.2.2   Voids in metal films

If there is a large concentration of non-equilibrium vacancies in a solid then they may diffuse and collect into complexes to form voids. We will call these voids of vacancy diffusion origin [4.6]. They are a three-dimensional collection of vacancies and may be considered as bubbles within the solid. They take on the equilibrium shape, which is usually spherical or cylindrical, and they are distributed uniformly throughout the film. Defects produced by other means, for example by contamination of the substrate, can take any form.

It is possible to consider a thin film, which has not been subjected to any annealing or natural ageing, as a layer containing a strongly saturated solution of vacancies on a foreign substrate. Such a film has an increased free energy because of the excess of non-equilibrium vacancies. This excess density of vacancies can decrease, to bring this film-substrate system into thermodynamic equilibrium, by the flow of vacancies to different sinks, where they can be annihilated or can collect together into stable complexes leading to the formation of voids.

These voids of vacancy diffusion origin can be formed in films during their storage and use, even at room temperature. The density and size of such voids can increase during the ageing of a film over several months. At increased temperatures (annealing) the process of void formation is accelerated. After the elimination of these excess vacancies in a film, a quasi-equilibrium concentration of vacancies is reached which still considerably exceeds the equilibrium concentration of vacancies for an ideal crystal because of the internal stress. This quasi-equilibrium concentration of vacancies in a film is determined by the formula [4.4]

$$n_v = \frac{N_v}{N_a} = A_v \exp[-(u_v - \sigma V_v)/kT] \tag{4.5}$$

where A_v is the entropy coefficient, σ is the mechanical stress. V_v is the volume of vacancy formation and $E_v = u_v - \sigma V_v$ is the activation energy of vacancy formation. For a bulk metal $A_v \approx 1$, but for thin films the factor $A_v \gg 1$, which accounts for the high concentration of vacancies in a film. It is known that the quasi-equilibrium concentration of vacancies in thin films at room temperature can be 0.1–1 at.%, which exceeds the equilibrium concentration of vacancies in bulk samples by several orders of magnitude [4.1, 4.2]. This is explained by the influence of a number of factors, in particular by the high concentration of sources (and sinks) for vacancies in the volume of a film, surface effects and also the high level of micro-strains at local sites in a film [4.2].

Let us consider the influence of mechanical stresses and surface effects on the concentration of vacancies in a film. The tensile stresses reduce the energy of vacancy formation, u_v, by an amount $E_v = u_v - \sigma V_v$, and the concentration of vacancies in a film is determined by the expression:

$$n_v = n_v^0 \exp(\sigma V_v/kT) \tag{4.6}$$

where n_v^0 is the concentration of vacancies in the film in the absence of mechanical stresses. Taking $\sigma = 10^9$ Pa, $V_v = 2 \times 10^{-29}$ m^3 at $T = 300$ K we obtain a value for the concentration of vacancies in a film in the presence of mechanical stresses as $n_v \approx 10^2 n_v^0$.

The equilibrium concentration of vacancies near a spherical micro-void is thus larger than that of a perfect film (without voids). The void may be considered as a bubble which evaporates vacancies into the condensate to enclose itself within a zone of excess vacancies, the concentration, n_v, of which is determined by the Frenkel formula:

$$n_v = n_v^0 \exp \frac{2\sigma_s V_v}{r_p kT} \tag{4.7}$$

where r_p is the radius of the void and σ_s is the surface energy. Palatnik with co-authors [4.2] showed that a spherical void is stable only if the number of vacancies in it exceeds $\sim 2 \times 10^3$. Smaller voids will dissolve into vacancies. The diameter of a void of this critical size is about 1 nm. For such sizes the thermodynamic description can still be used. If a void is larger than the critical size, then quasi-equilibrium conditions exist between the void and the bulk with the number of vacancies leaving the void in unit time balanced by the number of vacancies arriving from the bulk. Let us evaluate the concentration of excess vacancies near a micro-void at room temperature for an Al film using formula 4.7. Assuming that $r_p = 0.5$ nm, $\sigma_s = 1$ J m^{-2}, $V_v = 2 \times 10^{-29}$ m^3, we obtain $n_v = 2.5 \times 10^8 n_v^0$. For a micro-void with radius $r_p = 1$ nm, $n_v = 1.5 \times 10^4 n_v^0$. These calculations show that the concentration of vacancies in a film containing many small micro-voids can considerably exceed the equilibrium concentration of vacancies for an ideal crystal.

A real film contains voids of various sizes and shapes. A statistical description of such a pore system is not easy to make and the sizes of voids determined experimentally by different methods have not always agreed. For an aluminium film condensed in vacuum the sizes of micro-voids and their statistical size distribution were determined by Palatnik with co-authors [4.6]. With an overall volume density of voids of about 0.6% most of the voids (about 0.5% of the volume of a film) have a size $r_p \leq 4$ nm. This result confirms the existence in a film of small voids, which can considerably increase the concentration of vacancies in the film in agreement with equation 4.7.

The size of voids in bulk solids can be compared with the size of the crystallites [4.2]. Macro-voids refer to voids that have a size considerably larger than the size of the crystallites. For metal films such voids usually have a diameter $d_p \geq 1$ μm and can be seen visually with the help of an optical microscope. Micro-voids refer to voids which have sizes comparable with the sizes of the crystallites, and submicro-voids have sizes which are less than the elements of the structure and consequently can occur inside grains or within their boundaries. For metal films we shall use micro- and submicro-voids to refer to closed voids with a size less than the thickness of the film. Macro-, micro- and submicro-voids are defects that are typical of metal film structures. They arise because of the details of the condensation conditions of the metal on to the substrate. Stable submicro-voids can act as a source for the nucleation of macro-voids of vacancy diffusion origin

[4.6, 4.7]. Under certain conditions non-equilibrium vacancies can condense to form cavities on the surface of the film.

The thermodynamic conditions for the formation of voids or cavities and the kinetics of their growth depend on the details of the mechanical stresses in the film. Let us consider the process for the formation and growth of open voids in a metal film using the thermodynamic theory of heterogeneous condensation on surfaces developed by Volmer and Weber [4.8] and applied by Kurov and Zhil'rov to the kinetics of formation of open cylindrical voids by vacancy diffusion [4.7].

4.3 Kinetics of formation of open voids in thin films

As a model we will select an open cylindrical void (macro-void) of radius r_p in a sheet with a thickness h (Figure 4.2). We will assume that the open void forms as the result of heterogeneous condensation from a strongly saturated solution of vacancies within the film volume. We also assume that they diffuse along the surface of the film and coalesce into a complex.

Let us define $\Delta\Phi$ as the change in the thermodynamic potential of the film-substrate system due to the formation of an open cylindrical void with a radius r_p at temperature T; σ_{fv} is the surface energy of the film-vacuum boundary [J m^{-2}]; σ_{sv} is the surface energy of the substrate-vacuum boundary and σ_{fs} is the surface energy of the film-substrate boundary.

For $\Delta\Phi$ we have:

$$\Delta\Phi = \Phi - \Phi_0 \tag{4.8}$$

where Φ_0, Φ are the thermodynamic potentials of the film-substrate system before and after the formation of an open void.

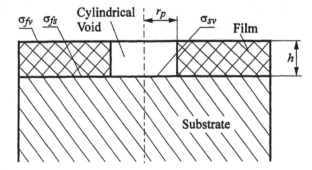

Figure 4.2 Diagram of an open cylindrical void in a thin film.

Let us write an expression for the change in the thermodynamic potential $\Delta\Phi$ allowing for the changes in the surface and volume energies of the film-substrate system:

$$\Delta\Phi = \pi r_p^2(\sigma_{sv} - \sigma_{fs} - \sigma_{fv}) + 2\pi r_p h \sigma_{fv} - \pi r_p^2 h \Delta\Phi_v \qquad (4.9)$$

where $\Delta\Phi_v$ is the change in the thermodynamic potential of the vacancy solution after the formation of a void of unit volume [J m^{-3}].

This can be expressed through the non-equilibrium concentration of vacancies as:

$$\Delta\Phi_v = \frac{kT}{V_v}\ln\frac{n_v}{n_v^0} \qquad (4.10)$$

where $n_v \geq n_v^0$ and $\Delta\Phi_v \geq 0$.

The angle of wetting for a cylindrical void in accordance with the second law of Laplace's capillarity is equal to $\pi/2$ and we have $\sigma_{fs} = \sigma_{sv}$. This gives for $\Delta\Phi$, from expression 4.9:

$$\Delta\Phi = 2\pi r_p h \sigma_{fv} - \pi r_p^2(\sigma_{fv} + h\Delta\Phi_v) \qquad (4.11)$$

This expression for $\Delta\Phi$ consists of two terms. The first term is positive and is proportional to the radius r_p, the second term is negative and proportional to r_p^2.

For a small void radius the second term is smaller than the first, so that $\Delta\Phi > 0$ and the process of void formation is not thermodynamically favourable. However for a void radius greater than that of the critical value at the potential maximum, the second term is larger than the first one, and the growth of a void is thermodynamically advantageous. Thus as the void radius increases the thermodynamic potential Φ passes through a maximum at the critical void radius $r_p = r_{pcr}$. The void size can reach the critical radius if there is a local fluctuation of the vacancy concentration or some other nucleation process.

Let us find the critical radius, r_{pcr}, of an open void where there is dynamic equilibrium between the evaporation and condensation of vacancies. The magnitude of r_{pcr} is found from the condition $\frac{\partial\Delta\Phi}{\partial r_p}\big|_h = 0$. After differentiating equation 4.11 we obtain:

$$\frac{\partial\Delta\Phi}{\partial r_p}\big|_h = 2\pi h \sigma_{fv} - 2\pi r_p(\sigma_{fv} + h\Delta\Phi_v) = 0 \qquad (4.12)$$

and hence:

$$r_{pcr} = h\sigma_{fv}\big/(\sigma_{fv} + h\Delta\Phi_v) \qquad (4.13)$$

Therefore, there is the size of a void for which any further growth reduces the thermodynamic potential of the system. Below this radius the void decays into vacancies. By substituting this expression into equation 4.11 we obtain an expression for the change in the thermodynamic potential of the system due to the formation of a void of the critical radius. This is also the energy of formation of the critical nucleus of an open cylindrical void:

$$\Delta \Phi_{cr} = \Delta \Phi(r_{pcr}) = \pi \sigma_{fv} h r_{pcr} = \frac{\pi h^2 \sigma_{fv}}{\sigma_{fv} + h \Delta \Phi_v} \qquad (4.14)$$

The equilibrium concentration per unit area of voids with the critical radius is determined by the expression [4.6]:

$$n_{pcr} = h n_v \exp(-\Delta \Phi_{cr}/kT) \qquad (4.15)$$

Here $\exp(-\Delta \Phi_{cr}/kT)$ is the Boltzmann factor defining the probability of energy fluctuation of magnitude $\Delta \Phi_{cr}$ because of the aggregation of vacancies. The product $h n_v$ gives the surface concentration of vacancies.

Let us define a density of super-critical open voids n_{po}, i.e. the number of voids in unit film surface area that are larger than voids of the critical radius. Their growth occurs due to the diffusion of vacancies along the film. We have [4.6]:

$$n_{po} = n_{pcr} \xi_v Z \tau_p \qquad (4.16)$$

where ξ_v is the number of vacancies colliding with the surface of a void of critical radius in unit of time; Z is Zel'dovich's non-equilibrium factor ($Z \leq 1$) and τ_p is the time for void formation. We will assume that the concentration of non-equilibrium vacancies remains constant with time.

The probability per unit time of a vacancy moving from one position of equilibrium into another is determined by the expression $f_0 \exp(-U_{vm}/kT)$ where f_0 is the frequency of thermal oscillations of an atom and U_{vm} is the activation energy of vacancy migration. For the value of ξ_v we have the expression:

$$\xi_v = V_{ef} n_v f_0 \exp(-U_{vm}/kT) \qquad (4.17)$$

where V_{ef} is the effective volume of a cylindrical annular ring of critical radius in a film of thickness h:

$$V_{ef} = \pi(r_{pcr} + a)^2 h - \pi r_{pcr}^2 h \approx 2\pi r_{pcr} a h \qquad (4.18)$$

Here a is the crystal lattice constant ($r_{pcr} \gg a$) and $2(r_{pcr} + a)$ is the external diameter and $2r_{pcr}$ is the inside diameter of the circular ring.

Using formulas 4.16, 4.17 and 4.18, we obtain an expression for the density of open voids:

$$n_{po} = 2\pi a f_0 r_{pcr} h^2 n_v^2 Z \tau_p \exp[-(\Delta\Phi_{cr} + U_{vm})/kT] \qquad (4.19)$$

If the film is under tensile stress the vacancy diffusivity is increased and the expression for the void density will become:

$$n_p = n_{po} \exp(\sigma V_{vm}/kT) \qquad (4.20)$$

where V_{vm} is the volume of vacancy migration which is the difference between the volumes of the strained and normal state of the crystal [4.4].

According to equation 4.20, an increase in the mechanical stress will increase the density of the macro-voids exponentially. This has been confirmed experimentally. With mechanical stress the vacancies diffuse along the film surface from large distances, several orders of magnitude larger than the thickness of the film [4.10], and then coalesce into voids through vacancy sinks.

An analysis of equation 4.19 shows that the density of voids in a film depends exponentially on the migration activation energy of the vacancies. Also, as U_{vm} increases, the density of voids decreases. To prepare films without voids it is necessary to increase the migration activation energy of the vacancies in the film material. The formation of voids in alloy films is not observed experimentally since the diffusion of vacancies is hampered. This occurs in Al/Si and Al/Cu alloys. Also the density of voids grows with an increase in the temperature.

Now we consider the dependence of the density of open voids on the thickness of a film at constant temperature and on the concentration of non-equilibrium vacancies. Let us consider the limit of very small thickness. Equation 4.13 shows that $r_{pcr} \sim h$, if $h \ll \sigma_{fv}/\Delta\Phi_v$, and from equation 4.14 we then have $\Delta\Phi_{cr} \sim h^2$. From equation 4.19 it follows that $n_{po} \sim h^3 \exp(-Ah^2)$ where A is some constant, which does not depend on h. For small thicknesses ($h \ll A^{-0.5}$), $\exp(-Ah^2) \approx 1$ so that equation 4.19 becomes $n_{po} \sim h^3$. That is, at small thicknesses the density of open voids grows as the cube of the thickness.

Let us now consider the limit of large thickness. For $h \gg \sigma_{fv}/\Delta\Phi_v$, equations 4.13 and 4.14 give $r_{pcr} = $ constant and $\Delta\Phi_{cr} \sim h$. In this case from equation 4.19 we have $n_{po} \sim h^2 \exp(-Bh)$ where B is a constant. From here it is easy to see that the density of open voids decreases exponentially as the film thickness increases:

$$n_{po} \sim \exp(-Bh) \qquad (4.21)$$

From these limiting cases it follows that for intermediate film thicknesses the density of voids has a maximum.

Let us now give the results of experimental studies of macroscopic voids in thin films [4.9]. The density of open macroscopic voids (diameter 1–3 μm) was investigated in films of aluminium and chromium, with thicknesses $h = 40$–200 nm, deposited on glass. Their technique used a metallurgical microscope with transmitted light.

The image was observed as a dark field on which bright points are visible. The open circular voids of vacancy diffusion origin have an image in the form of a circle. These are cylindrical voids which are distributed, as a rule, uniformly over the film surface. An irregular distribution is characteristic of voids which were nucleated by contamination on the substrate, which usually indicates sub-standard cleaning of the substrate, macroscopic surface imperfections such as cracks and scratches, or a non-uniform distribution of temperature along the substrate during condensation. Such voids also have an irregular form, rough edges and their size may sometimes exceed the size of the circular voids.

For a quantitative evaluation of the porosity of films Kurov with co-authors [4.9] used a microscope eyepiece with a magnification $\times 600$ with a measuring grid which had, in the field of view, 128 small squares each with an area of 5×5 μm^2. The void density, n_p, of the film was obtained by counting the number of voids in each of the 128 units. The density of voids was determined from the number of holes with a diameter greater than 1–3 μm per unit area:

$$n_p = \frac{N_p}{A_0} \qquad (4.22)$$

where N_p is the number of voids in the field of view of the eyepiece and A_0 is the area of view of the eyepiece.

Figure 4.3 represents the dependence of the density of open circular voids as a function of the thickness in aluminium films produced with the same condensation velocity, $w_c = 3$ nms^{-1} but for different substrate temperatures ($T_{C1} < T_{C2} < T_{C3}$) during the film deposition [4.9]. It can be seen that all the curves have a maximum. The magnitude of the maximum of the void density and its position change with the condensation temperature. This agrees qualitatively with the theoretical conclusions that the density of voids increases with an increase in the condensation temperature and that the position of the maximum moves to larger thicknesses.

The formation of macro-voids of diffusion vacancy origin in films is enhanced by tensile thermal stresses. This is confirmed by the results of a measurement of the density of macro-voids in aluminium films deposited on substrates with various coefficients of thermal expansion [4.11]. These experiments provided a check on the role of thermal stress (see Section 4.4.4) in the formation of macro-voids. Films with a thickness $h \approx 100$ nm were deposited in vacuum simultaneously on to substrates of three materials: glass, fused quartz and single crystal sodium chloride with orientation (100),

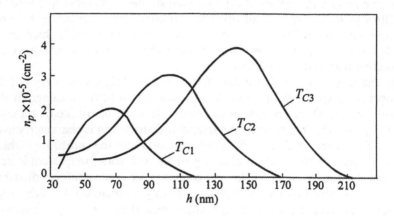

Figure 4.3 Thickness dependence of the open void density in aluminium films under different condensation temperatures $T_{C1} < T_{C2} < T_{C3}$ [4.9].

at a condensation temperature $T_c = 400\,\mathrm{K}$ and a condensation rate $w_c = 5\,\mathrm{nm\ s^{-1}}$. Immediately after condensation the samples were heated in the vacuum chamber for 1200 s at a temperature equal to the temperature of condensation. The measurements were made after cooling to room temperature. In films deposited on the fused quartz substrates the density of open voids was high: $n_p = 10^5\,\mathrm{cm^{-2}}$. The films on the glass substrates had considerably fewer voids and in the films deposited on sodium chloride voids were completely absent. These results are explained by the differences in the magnitude and sign of the thermal stresses in the films deposited on the different substrates.

The greatest tensile thermal stresses arise in aluminium films on the fused quartz. In these films the porosity is higher than in films on glass. In films deposited on sodium chloride the thermal stresses are compressive. Open voids were not observed in such films. Therefore, it is possible to conclude that tensile thermal stresses in films promote the formation of voids originating from vacancy diffusion. A high density of open macro-voids in metal films is observed under high tensile stress for Cr films with $\sigma \geq 3 \times 10^8$ Pa and for Al films with $\sigma \geq 1 \times 10^8$ Pa.

It should be noted that with improvements in the manufacturing technology of metal films the concentration of vacancies in metal films will become lower. So in films made on modern equipment open circular pores may not occur. However micro-voids, with sizes about the size of crystallites, and submicro-voids, with sizes much less than any element of the structure, of vacancies diffusion origin will be formed in films during their storage and operation. This occurs for Al and Cr films deposited in an atmosphere of argon (see Section 4.6) and for Al/Cu alloy films. The kinetics of the origin of spherical voids in thin films is largely similar to the kinetics of the open cylindrical voids considered in this section.

4.4 The internal stresses in thin films

The origins of the macro-stresses that occur in films that are firmly attached to the substrate have been described in detail in the literature [4.1, 4.2]. We give here a brief classification of these mechanical stresses based on the physical mechanisms that cause them.

4.4.1 Structural stresses

When a film is deposited on a substrate, the condensate is saturated with the various defects of a crystal lattice (excess vacancies, bivacancies, dislocations etc.). The film can contain micro-voids included in its volume. Films prepared by evaporation-condensation in a vacuum or by ion sputter deposition are usually formed under conditions that correspond to crystallisation from a strongly saturated vapour. Then excess vacancies are captured by the condensate during the formation of the film. Because of the rather small mobility of such vacancies inside the condensate they have no time to move to the front of the condensation and hence leave the volume of the film. It has been shown by Palatnik with co-authors [4.2] that the average speed of the vacancies is one to two orders of magnitude smaller than the growth rate of the film condensate w_c. Thus the concentration of the captured vacancies in a film can reach ~ 0.1–1 at.%. At a low condensation temperature it is possible for the excess vacancies to form vacancy accumulations, which act as sinks [4.2]. During the formation of a film, and also under subsequent heat treatment and storage, there is a gradual decrease in the number of defects in a crystal lattice by the motion of vacancies to sinks, the coalescence of separate defects etc. These processes reduce the free energy of the film. All this results in a change in the specific volume of the film and, as a consequence, in the size of the structural stresses. Stresses can arise because a decrease in the vacancy density in the film volume is energetically favourable since the elastic energy of these stresses is some orders of magnitude smaller than the total energy of the initial quantity of non-equilibrium vacancies.

The effect of dislocations on the change of the specific volume of a crystal is small in comparison with that of vacancies and impurity atoms. Therefore the change in the stresses within a film which arise from a change in the concentration of dislocations is less than that from the change in the concentration of excess vacancies.

In metals, as a rule, the concentration of interstitial atoms is negligibly small compared with the concentration of vacancies. Thus the increase in the volume of a crystal by an amount ΔV can be presented as a linear function of the atomic concentration of vacancies according to Schottky:

$$\frac{\Delta V}{V_0} = A_1 n_v \qquad (4.23)$$

where A_1 is a constant and V_0 is the volume of the crystal with a perfect lattice, without vacancies. In metals the volume of an isolated vacancy is found to be less than the volume filled by one atom in a perfect lattice. In particular, for a face-centred cubic lattice $A_1 = 0.7$: for a body-centred lattice $A_1 = 0.5$ [4.5].

The different methods of removing excess vacancies can influence the sign of the consequent macro-stresses differently:

1   When the vacancies move to an outside surface, into a macro-void within the film or to the line of an edge dislocation, the specific volume of the condensate decreases and tensile stresses will arise within the film;
2   When the vacancies coalesce into micro-voids the specific volume of the condensate grows, since the volume of a micro-void is equal to the sum of the volumes of the atoms while the volume of the same number of isolated vacancies is approximately half the size. Therefore coalescence of vacancies into a micro-void produces compressive stresses.

Thus the structural stresses can be either tensile or compressive. Condensates closer to equilibrium have a smaller concentration of micro-defects and a more ordered structure. They also have lower structural macro-stresses.

4.4.2   Phase stresses

Phase stresses arise because of the phase heterogeneity of a film. Under some deposition conditions the metal is initially condensed in an amorphous state or crystallised into various metastable modifications (see Section 3.4.1). At this stage of film formation there are no phase stresses. When the metastable phase turns into a stable one, phase stresses arise if the two phases have different densities. The subsequent processes of film crystallisation or recrystallisation are then accompanied by a change in the specific volume that also produces macro-stresses which can be called phase stresses. These phase stresses can be either tensile or compressive.

4.4.3   Physical and chemical stresses

Physical and chemical stresses result when different chemical compounds form within a film or dopant gas atoms are included. The basic processes that result in physical-chemical stresses are; film oxidation which occurs when there are residual gases in the chamber during the preparation of the film or the film is deposited in an ambient atmosphere, and the inclusion of impurity atoms in the lattice during the condensation of the film. As the specific volume of oxide is more than the specific volume of the metal, and the volume of the included molecules of gases is more than the volume of the metal atoms, both of the above processes result in an increase in the specific volume of the condensate. Therefore the physical-chemical stresses are always compressive.

These kinds of mechanical stresses in films: structural stresses, phase stresses, physical and chemical stresses are called self stresses. Under real conditions all the above mechanisms for the origin of self stresses in films can act simultaneously. Depending on the physical-technological conditions of the film preparation only one may be observed or the summation of stresses of the same sign or their cancellation because of the different signs of the various kinds of stresses. If the conditions of film preparation are changed, the contributions of the stresses due to the various mechanisms of formation also change.

As well as the self stresses there are also thermal stresses.

4.4.4   Thermal stresses

The condensation of metal films is usually carried out on a heated substrate at a temperature T_c. The reduction of the film temperature, from the condensation temperature down to room temperature, after the end of the condensation process, causes a change in the linear sizes of both the film and the substrate. As the coefficients of thermal expansion (CTE) of the film and substrate materials are always different, thermal stresses arise on cooling. Similarly, when the film temperature is increased above the condensation temperature, thermal stresses arise due to the heating.

For a film that is strongly bonded to its substrate the following expression for the relative deformation of a film is obtained:

$$\varepsilon_T = (\beta_f - \beta_{sub})(T_c - T) \tag{4.24}$$

where β_f and β_{sub} are the linear coefficients of thermal expansion of the materials of the film and substrate respectively, and $T_c - T = \Delta T$ is the difference between the temperature of condensation and the temperature of the stress measurement. The resulting thermal stresses, assuming biaxial symmetric stress condition, are determined by equations 4.2 and 4.24 and are given by:

$$\sigma_T = \frac{E_f}{1 - \nu_f}(\beta_f - \beta_{sub})\Delta T \tag{4.25}$$

For the case $\beta_f > \beta_{sub}$ and $\Delta T > 0$ the thermal stresses, $\sigma_T > 0$ so that they are tensile stresses.

Equation 4.25 holds if the deformation in the film does not exceed its elastic limit so that Hooke's law (equation 4.2) is still valid. The thermal stresses are frequently large and can be of either sign. However they can be estimated using formula 4.25 with good accuracy, and checked.

For aluminium films on a cover glass the thermal stresses can reach a value $\sigma_T \approx 1.5 \times 10^8$ Pa, and for chromium films $\sigma_T \approx 0.6 \times 10^8$ Pa. For Cr films the thermal stresses are thus less than for Al films. This is explained by a smaller difference in the temperature coefficients of expansion for chromium and glass than for aluminium and glass.

The thermal stresses add algebraically to the self stresses and are then usually called the total stresses or macro-stresses given by:

$$\sigma = \sigma_0 + \sigma_T \tag{4.26}$$

4.5   The measurement of internal stresses in films

Many methods are used for the measurement of the internal macro-stresses in films. Let us consider the most basic of them: X-ray and electron microscope stress measurements, the deformation of a circular substrate and the curvature of an elastic beam substrate. In all these methods it is assumed that the stresses are homogeneous within the thickness of the film. However, real stresses in a film are usually inhomogeneous through the thickness, which explains the twisting of films after they separate from the substrate.

4.5.1   X-ray and electron diffraction methods

The internal stresses in a film can be determined by measuring the change in the lattice constant in the plane of the film with the help of X-ray or electron diffraction patterns. The X-ray diffraction method is preferable because of its higher resolving power [4.12, 4.13]. In these methods the stresses are calculated using the formula:

$$\sigma = \frac{E_f}{1 - \nu_f} \left(\frac{a - a_0}{a_0} \right) \tag{4.27}$$

where a_0 and a are the lattice constants of the bulk sample and the deformed film respectively.

4.5.2   Method of deformation of a circular substrate

To determine the substrate deformation after the deposition of a film the change in the interference pattern formed in the gap between a circular substrate and a flat optical glass is used. As substrates are not absolutely flat, it is necessary to determine the profile of the substrate again after removal of the film, using an appropriate chemical fluid or after reactive ion etch, and that profile can be used as a reference. The curvature of a substrate can also be determined before the deposition of the film. The deformation of a circular plate allows the measurement of the anisotropy of the stresses in the plane of a film. The stresses in a film are determined from the measured deformation of the plate using standard elasticity theory. Under equilibrium conditions we have a relation between the isotropic stresses, σ, and the resulting strain of a substrate [4.1]:

$$\sigma = \frac{E_{sub}\, h^2_{sub}}{6(1 - \nu_{sub})R_{sub}h} \tag{4.28}$$

where E_{sub} and ν_{sub} are the Young's modulus and the Poisson ratio of the substrate respectively; h_{sub} and h are the thicknesses of the substrate and film respectively and R_{sub} is the radius of curvature of the substrate produced by the stresses in the film.

As the stresses in the film are calculated only from the observable strain in the substrate, the elastic constants of the film do not appear in the calculation, in this approximation.

4.5.3 *Method of elastic beam substrate curvature*

The internal mechanical stresses in a film are determined from the deviation of the free end of a cantilever substrate, fixed at one end, before and after etching the film, or before and after film deposition. Sometimes the deviation of the substrate is measured immediately after the film deposition in the vacuum chamber. This allows the measurement of the stresses in the film during the condensation process and then during the annealing and heat treatment of the film. Then during the measurement of the stresses in the vacuum chamber the film is not oxidised by heating.

Figure 4.4 shows the curving of the cantilever substrate with a deposited film. After etching the film the substrate will return to the "zero", non-stressed, position and the free extremity will deviate by a distance δ.

The substrate is selected with a length 5–10 times greater than its breadth. The thickness of the film is much less than the thickness of the substrate $h \ll h_{sub}$. The adhesion between the film and the substrate should be strong. The measurement of the deviation of the free extremity of the substrate is made with the help of an optical microscope focussed on the end face of the substrate. A micrometric object-glass with a graticule is usually used.

Elasticity theory allows us to calculate the stresses in a film as a function of the deviation, δ, of the free end of the cantilever substrate. For small deviations the stresses in a film are calculated using Stony's formula with allowance made for the Poisson ratio:

$$\sigma = \frac{E_{sub}\, h^2_{sub}\delta}{3hL^2(1 - \nu_{sub})} \tag{4.29}$$

where E_{sub} and ν_{sub} are Young's modulus and the Poisson ratio of the substrate respectively; h_{sub} and h are the thicknesses of the substrate and

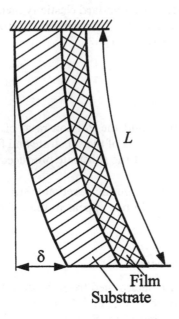

Figure 4.4 The curving of the substrate fixed as a cantilever under the action of the internal mechanical stresses in the film.

the film respectively; L is the length of the cantilever. In the above theory it is assumed that the displacements under bending are small in comparison with the thickness of the plate.

As the stresses in the film are calculated for an observable deformation of the substrate, the same as in the method of the deformation of a circular plate, the elastic constants of the film again do not appear in the Stony formula (equation 4.29). It should be noticed that the inclusion of the Poisson ratio in equation 4.29 increases the stresses by $1/(1 - \nu_{sub})$ ≈ 1.5 times.

Equation 4.29 determines the average macro-stresses within the thickness of the film. The elastic substrate bending method is the most sensitive for the measurement of the internal stresses in films. For a substrate of size $60 \times 4 \times 0.15$ mm and a film thickness of $h = 100$ nm the error in the measurement of the stresses is less than $\delta\sigma = \pm5\%$ for $|\sigma| > 2 \times 10^8$ Pa, $\delta\sigma = 10\%$ for 10^8 Pa $\leq |\sigma| \leq 2 \times 10^8$ Pa, and $\delta\sigma = \pm25\%$ for $|\sigma| < 10^8$ Pa.

The self stress, σ_o, is defined by equation 4.26 by subtracting any thermal stress, σ_T, calculated using equation 4.25, from the total measured stress.

For the calculation of thermal stresses using equation 4.25 it is necessary to know Young's modulus and the Poisson ratio for the metal film. Usually these are assumed to be the same as for bulk metals. The mechanical properties of some metals are presented in Table 4.1 [4.14].

Table 4.1 Physical and mechanical properties of metals [4.14]

Metal	Density [g cm^{-3}]	CTE [10^{-6} K^{-1}]	Young's modulus [GPa]	Poisson ratio, v	Yield point [GPa]	Strain limit [GPa]
Al	2.7	25	65	0.35	0.17	–
W	19.2	4.4	360	–	4.0	1.2
Au	19.3	14	82	0.42	–	0.14
Cu	8.9	16	120	0.34	–	0.22
Mo	10.0	5.4	350	0.32	2.1	0.7–1.2
Ni	8.9	13	210	0.30	–	0.45
Pd	12.0	12	176	–	–	0.20
Pt	21.4	9	170	0.39	–	0.15
Ag	10.5	19	80	0.38	–	0.18
Ta	16.6	6.6	190	–	–	0.50
Ti	7.0	8	90	–	–	0.30
Cr	7.0	7	68	0.13	–	–

Table 4.2 Physical and mechanical properties of semiconductor substrates [4.14]

Material	Orientation	Young's modulus [GPa]	Poisson ratio, v	CTE [10^{-6} K^{-1}]
Si	(100)	130–169	0.065–0.279	3.0–4.5
	(130)	130–174	0.092–0.304	
	(111)	169	0.262	
Ge	(100)	104–138	0.028–0.270	6.0
	(130)	104–143	0.060–0.296	
	(111)	138	0.249	
GaAs	(100)	85–121	0.021–0.312	5.9
	(130)	85–127	0.061–0.349	
	(111)	121	0.303	

It has been found that the coefficient of thermal expansion (CTE) of metal films with a thickness 40–60 nm, obtained by condensation in a vacuum, does not differ significantly from that of bulk materials. The coefficient of thermal expansion of a glass substrate is usually accepted to be equal to $\beta_{gl} = (7-9) \times 10^{-6}$ K^{-1} and for silicon wafers $\beta_{si} = (3-4.5) \times 10^{-6}$ K^{-1}. Silicon oxide, SiO_2, expands little so that at $T = 297$ K its coefficient is equal to 0.5×10^{-6} K^{-1}. In Table 4.2 the mechanical properties of some semiconductor substrates are given [4.14].

4.6   Experimental data on mechanical stresses in films

The dependence of the macro-stresses in thin films on the film thickness, the condensation rate and the temperature have been measured. The magnitude of the internal stresses are also influenced by the annealing and thermo-cycling of films in vacuum.

Figure 4.5 shows the dependence of the average self macro-stresses for copper and silver films on the thickness of the film [4.1]. It can be seen from the figure that no stresses appear up to a thickness of some tens of nanometers, while the film is discontinuous (island like). With an increase in the thickness the islands join and internal stresses appear. They become significant at a critical thickness $h_{cr} \approx 20\text{--}30\,nm$. Then the stresses increase sharply up to a defined level.

At a film thickness of 50–100 nm the formation of a continuous film is complete. The stresses reach a maximum value frequently exceeding the limit of the bulk material. After a further increase in the film thickness the size of the macro-stress decreases. This is because almost all of the stresses are located near the substrate in a thin layer of the film with a thickness $h \leq 100\,nm$. Therefore the average macro-stress decreases with an increase in the film thickness.

The dependence of the self stresses in metal films on the temperature of condensation has been measured. At condensation temperatures up to 400–500 K there are frequently tensile self stresses caused by structural stresses. At higher temperatures of condensation the stresses become compressive. The origin of these compressive stresses is explained by the oxidation of the film along the grain boundaries from the atmosphere of the residual gases. The analysis thus reduces to the origin of the compressive physical and chemical stresses. For example, compressive stresses are observed in copper films if the pressure of the residual gases in the chamber during the deposition exceeds $10^{-2}\,Pa$. Electron diffraction patterns of such film samples show diffraction rings appropriate to an oxide of copper Cu_2O [4.1].

The condensation rate has a significant influence on the stresses. If it is increased the grains of the film are smaller, with a low impurity content, as the number of residual gas molecules introduced into the film is reduced.

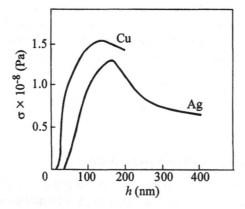

Figure 4.5 Thickness dependence of the macro-stresses in films of copper and silver.

Similarly an increase in the condensation rate increases the number of vacancies captured by the film, so they have no time to go out from the film volume to the condensation front and then leave the film. Further, during the ageing or annealing of the film there is diffusion of vacancies to various sinks. The loss of vacancies to the surface of the film or into open macrovoids decreases the specific volume of the condensate and therefore leads to an increase in the tensile stress. For example, for aluminium films of thickness 100 nm deposited on a cover glass or an oxidised silicon wafer by thermal evaporation in a vacuum (with a pressure of residual gases in the chamber less than 10^{-3} Pa) an increase in the condensation velocity of 4 up to 10 nm s^{-1} results in an increase in the macro-stresses by approximately three times from 1×10^8 Pa up to 3×10^8 Pa.

The annealing and thermocycling of films in a vacuum produces irreversible changes in the structure of films. After a low temperature anneal, where the temperature of annealing, T_{ann}, is lower than the recrystallisation temperature, T_{rec}, the number of non-equilibrium point imperfections in the lattice decreases and dislocation flow takes place. After a high-temperature anneal ($T_{ann} > T_{rec}$) a decrease in the number of disoriented grains occurs because of the modification of the grain boundaries and the growth of the size of the grains; that is recrystallisation takes place.

Figure 4.6 shows the typical dependencies of the macro-stresses in Al films, deposited in vacuum and an atmosphere of argon, with temperature during thermocycling [4.15]. The conditions for thermocycling were selected as the following: the heating was carried out over 1800 s up to a temperature of 450 K, the samples were kept at this temperature for 600 s and were then cooled during 1800 s down to 320 K. Then the vacuum chamber was

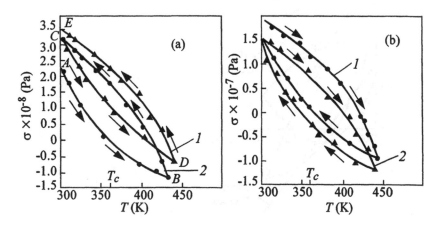

Figure 4.6 The dependence of the strain, σ, on temperature in Al films obtained by evaporation in (a) vacuum and (b) an atmosphere of an argon ($P_{Ar} = 10^{-1}$ Pa): curve 1 – first thermocycle; curve 2 – second thermocycle [4.15].

depressurised. The samples, which were subjected to two cycles, were cooled to approximately room temperature after the first thermocycle, and then the second thermocycle was carried out.

Note that the second (or third) cycle of heating and cooling modifies the internal stresses only if the maximum temperature of the previous cycle is exceeded. Then further irreversible modifications of the macro-stresses in the films may be observed. This observation indicates that there are imperfections in films with different annealing energies and the imperfections with smaller activation energy are eliminated at the lower temperature.

It can be seen in Figure 4.6a that the cooling curve for the stresses in films deposited in vacuum is situated above the equivalent heating curve. This verifies that the macro-stresses increase irreversibly on thermocycling. For films produced with a high enough pressure of argon in the chamber ($P_{Ar} > 10^{-2}$ Pa) the cooling curve passes below the heating curve. That is, after thermocycling the level of tensile stresses in a film is reduced (Figure 4.6b).

Figure 4.7 represents the dependence of the self stresses in aluminium films deposited in vacuum on the temperature of the first thermocycle and the thermal stresses [4.11]. The self stresses are obtained by the subtraction of the thermal stresses, calculated using equation 4.25, from the total macro-stresses. Research into the behaviour of voids after thermocycling of vacuum deposited aluminium films has revealed a large number of voids with

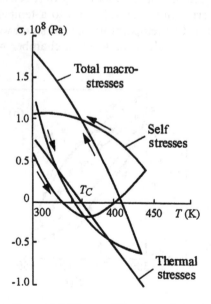

Figure 4.7 The modification of the internal mechanical stresses in an aluminium film after annealing during the first thermocycle showing the self stresses, $\sigma_0 = \sigma - \sigma_T$; the total macro-stresses, σ, and the thermal stresses, σ_T [4.11].

a diameter of about 1 μm up to a thermocycling that corresponds to a level of mechanical stresses in the film of $\sigma \approx 2 \times 10^8$ Pa (point A in Figure 4.6a).

When a film is heated to a temperature T $\approx$ 440 K, as well as a decrease in the size of the stresses and their change towards compressive stresses, voids are observed to disappear with the simultaneous formation of pits on the film surface (point B in Figure 4.6a).

On the subsequent cooling of the sample, the compressive stresses change into tensile as some of the pits formed during the heating turn into open voids with a diameter 1–2 μm. With a further decrease in the temperature the voids increase in size. At a temperature of 400 K some voids achieve sizes up to 3 μm or larger. An irreversible increase of the tensile stresses up to a value $\sigma \approx 3 \times 10^8$ Pa (point C in Figure 4.6a) is also observed. Points D and E in Figure 4.6 correspond to the second thermocycle. It is possible to explain the observed behaviour by the annealing of the vacancies captured in the film and the loss of vacancies into macro-voids. The amalgamation of vacancies to form open macro-voids results in a decrease in the specific volume of the film [4.2] and, therefore, an increase in the structural macro-stresses (Figure 4.7).

In films of chromium and aluminium, deposited in an atmosphere of argon, open macro-voids of vacancy diffusion origin are not formed. These films also have a lower level of tensile stress. The decrease of the macro-stresses produced by annealing these films (Figure 4.6b) indicates that the non-equilibrium vacancies join into complexes of closed micro- and sub-micro-voids. For such amalgamations of vacancies the specific volume of the condensate is increased, since the volume of a micro-void is equal to the sum of the volumes of the atoms, while the volume of the same number of isolated vacancies is approximately half that size. Therefore this process produces a decrease in the tensile stress. This is explained by the fact that a metal film deposited in an atmosphere of argon at high enough pressure ($P_{Ar} > 0.1$ Pa for Al) contains fewer non-equilibrium vacancies and other imperfections of the crystal lattice, and hence has a more perfect structure, than a film deposited in a uncontrolled medium of residual gases. The lack of open macro-voids of vacancy diffusion origin in such films is explained by the following observation. Macro-voids in a film may arise due to the amalgamation of excess vacancies if their concentration is large. In the initial stage of the nucleation, micro-voids have a spherical form due to the mini-misation of the total free energy and they are within the film volume. Further growth of the voids occurs by the diffusion of vacancies towards the micro-void from the diffusion zone. When a micro-void reaches a diameter equal to the thickness of the film it becomes an open macro-void. However, if the concentration of non-equilibrium vacancies in a film is too small, macro-voids cannot be formed and the vacancies are integrated into small complexes and micro- and submicro-voids. This can explain the lack of macro-voids in the Al and Cr films deposited in argon. The amalgamation of vacancies into complexes or micro-voids can be energetically profitable because the number of broken bonds in the crystal lattice decreases. The formation of vacancy complexes promotes a decrease in the elastic strain energy of a film that is exhibited as a decrease in the tensile stresses after heat treatment of the films (Figure 4.6b).

During ageing of films, processes occur which produce a change in the specific volume that in turn results in a modification of the macro-stresses with time. If the decrease in the specific volume of a film is connected with the removal of surplus vacancies into macro-voids then the growth of macro-stresses will be observed.

Literature for Chapter 4

4.1   Hoffman, R.U., "Mechanical property of thin condensed films", *Physics of Thin Films*, Hass, G. and Thun, R. (eds), Academic Press, New York and London, 3 (1966) 211–273.

4.2   Palatnik, L.S., Fuks, M.Ya. and Kosevich, V.M., *Mekhanizm Obrazovaniya i Struktura Kondensirovannykh Plenok* (*Mechanism of fabrication and structure of condensed films*) (in Russian) Nauka, Moscow, 1972.

4.3   Timoshenko, C.P. and Gud'er, D., *Theory of Elasticity* (in Russian) Nauka, Moscow, 1975.

4.4   Girifalko, L.A., *Statistical Physics of Materials*, J. Wiley, New York, 1973.

4.5   Damask, A. and Dienes, J., *Point Defects in Metals*, Pergamon Press, New York, 1963.

4.6   Palatnik, L.S., Cheremskoy, P.G. and Fuks, M.Ya. *Pori v Plenkach* (*Voids in Films*) (in Russian), Metallyrgiya, Moscow, 1982.

4.7   Kurov, G.A. and Zhil'kov, E.A., "Macro-voids in thin films", *Mikroelektronika*, 1 (1972) 144–151.

4.8   Hirth, J.P. and Paund, G.M., *Condensation and Evaporation: Nucleation and Growth Kinetics*, MacMillan, New York, 1963.

4.9   Kurov, G.A., Markaryan, A.B. and Zhil'kov, E.A., "Macro-voids in thin metal films", *Mikroelektronika*, 2 (1972) 145–153.

4.10   Kurov, G.A., Zhil'kov, E.A. and Dubodel, B.M., "Macroscopic defects in thin metal films", *Sov. Phys. Dokl.*, 19 (1975) 772. (*Dokl. Akad. Nauk SSSR.* 219 (1974) 582–585.)

4.11   Kurov, G.A., Zhigal'skii, G.P. and Brilov, I.N., "Investigation of mechanical stress and voids in thin Al films", in *Fizika poluprovodnicov i mikroelectronika* (*Physics of Semiconductors and Microelectronics*) (in Russian), RRTI Publ., 1978, pp. 90–94.

4.12   Cohen, B.G. and Focht, M.W., "X-ray measurement of elastic strain and annealing in semiconductors", *Solid State Electron.*, 13 (1970) 105–112.

4.13   Louzon, T.J. and Spencer, T.H., "X-ray diffraction stress measurement in thin films", *Solid State Technol.*, 7 (1975) 25–28.

4.14   Zaharov, N.P. and Bagdasaryan, A.V., *Mechanical Phenomena in Integrated Structures* (in Russian), Radio i Svyaz', Moscow, 1992.

4.15   Kurov, G.A. and Brilov, I.N., "Influence of argon on some properties of thin aluminium films", *Fizika poluprovodnicov I mikroelectronica* (*Physics of Semiconductors and Microelectronics*) (in Russian), RRTI Publ., 1979, pp. 75–79.

5 Electrical noise in thin films and thin-film structures

5.1 Basic fluctuation theory

Fluctuations are the random deviations of the magnitude of quantities from their average values. The characteristics of these fluctuations can be very varied. It is difficult to find any quantity in physics, and generally in nature, whose numerical parameters do not have fluctuations.

The fluctuations of voltage and current, which are known as "noise", in electrical circuits determine the limiting sensitivity of the measurement of electrical signals. Noise reduces the useful information and may even make it impossible to understand. Therefore fluctuations in radio-physics are considered harmful and attempts are made to reduce them, especially in highly sensitive measuring devices.

On the other hand, the fluctuations contain valuable information about the dynamic behaviour of a system even when it is in equilibrium. In some systems only an understanding of the noise can determine the nature and mechanism of a particular phenomenon occurring.

In practice fluctuations give information about the unstable nature of systems and so their analysis gives information to improve a system. Thus at the engineering level the deficiencies of electronic devices, resistors, condensers, transistors, ICs, etc. are studied. It has been established that the devices, within a batch of nominally identical devices, which show greatest noise are the least reliable. Noise is a sensitive indicator of latent defects in electronic devices since little difference in other electro-physical properties can be observed. Some performance parameters of semiconductors, such as the current carrier lifetime and deep trap level parameters can be determined from the measurement and analysis of noise.

Fluctuations are measures of the physical parameters of macroscopic systems. By this we usually mean a system constructed from a large number of atoms and molecules. Many of the properties of such systems cannot be explained at an atomic level due to the interaction of the large number of atoms in these systems, perhaps 10^{10}–10^{20}, which would require the solution of a large number of equations describing the interactions between the atoms. Also because of the huge number of interacting atoms new macroscopic properties are observed which cannot be related to the properties of isolated, individual atoms or for a system consisting of several atoms. We will give an example.

Consider a closed vessel with one gas molecule moving inside it. This molecule collides with the walls of the vessel and transmits a certain momentum (impulse) depending on the conditions. If there are several molecules in the vessel, the momentum transferred to the walls per unit time will increase but it is hardly possible to speak here about a gas pressure since the transferred momentum per unit of time, per unit of area, will vary considerably with time. If the number of molecules in $1\,cm^{-3}$ reaches, for example, the magnitude $10^{10}\,cm^{-3}$, the momentum transferred to the wall surface in unit time will stabilise in the neighbourhood of some constant value. Thus the macroscopic behaviour of a gas in a vessel can be described by a quantity called the gas pressure, which in the case of only a few molecules has no sense.

Similarly it is possible to speak about both the gas density and temperature. The pressure, density and temperature of the gas describe the macroscopic properties of the system and they are the result of the superposition of a huge number of micro-processes.

Thus a considerable increase in the number of atoms and molecules in a system results in qualitatively new macroscopic properties of the system which behave in the same way as the random quantities. The fluctuations in the magnitudes of the physical or electrical quantities are the natural consequence of the atomic nature of the substance and the discreteness of the electrical charge. Any electrical phenomenon in solids can be described eventually by the motion of the current carriers and their scattering by various centres so that their motion becomes random and the velocities of the current carriers are continuously and randomly changed. Any macroscopic effect is measured after many microscopic collision and scattering events of particles in the system. So the physical parameters characterising this effect in the macro-system will be fluctuating values.

We see that the fluctuations are the characteristics of the physical parameters of the macro-system. In radio-physics the fluctuations of voltage and current in the various elements of integrated circuits and electronic devices are called "electrical noise" or simply "noise".

5.2   Statistical description of noise

5.2.1   *Noise power spectral density*

Electrical noise can be represented by the random sequence of impulses of voltage or current, following one after another in random intervals of the time. This process of random impulses is non-repeating. Nevertheless it is possible to speak about a spectrum of such a process which is the distribution of the power or intensity with frequency. To describe noise we introduce the concept of the noise power spectral density (PSD), or spectral density (SD), which is defined by the relation [5.1]:

$$S(f) = \lim_{\Delta f \to 0} \frac{\Delta P(f)}{\Delta f} \tag{5.1}$$

where $\Delta P(f)$ is the time-average power of the noise in a frequency band Δf at the measurement frequency f. The power spectral density has dimensions $W \ Hz^{-1}$. Generally the PSD is a function of frequency and is called the spectrum or the power spectrum.

The total power of the noise in the frequency band from f_1 to f_2 is equal to:

$$P = \int_{f_1}^{f_2} S(f) df \tag{5.2}$$

If the PSD of the noise is constant in the frequency band f_1 to f_2 passed by the circuit or the measuring instrument and is equal to S_0, then the total noise power in this frequency band is $P = S_0(f_2 - f_1)$.

In practice in the evaluation of the noise magnitude of any element or device in the radio spectrum the mean square average of the voltage, $\overline{\Delta U^2}$, in units of V^2 or the mean square average of the current, $\overline{\Delta I^2}$, in units of A^2 are usually measured. Thus the PSD of a noise is expressed in units of $V^2 \ Hz^{-1}$ or $A^2 \ Hz^{-1}$ and the spectral density of voltage fluctuations, $S_U(f)$, or current fluctuations, $S_I(f)$, are calculated using the formulas:

$$S_U(f) = \lim_{\Delta f \to 0} \frac{\overline{\Delta U^2}}{\Delta f}, \qquad S_I(f) = \lim_{\Delta f \to 0} \frac{\overline{\Delta I^2}}{\Delta f} \tag{5.3}$$

The line over the top of the quantities represents the time-averaging operation.

If there are several independent sources of noise in the linear elements of an electrical circuit, the full noise power for the circuit will be equal to the sum of the powers from each of the sources. Therefore the average power of the total noise voltage (or current) is determined from the sum of the average powers of all the separate noise sources. That is the noise voltages (currents) add quadratically. If there are two independent noise sources U_1 and U_2 in linear elements of an electrical circuit, the total mean average square of the noise voltage is

$$\overline{U^2} = \overline{U_1^2} + \overline{U_2^2} \tag{5.4}$$

5.2.2 The auto-correlation function and the Wiener-Khintchine theorem

The auto-correlation function is an important description of random process as well as the power spectral density. It shows the correlation of the random fluctuations with time delays.

Let $x(t)$ be a random function of time, a random process such as a voltage or current. Let us select two instants of time, t and $t+\tau$. The auto-correlation, or correlation, function, $K(\tau)$, is defined as the average with time of the product of the random values $x(t)$ and $x(t+\tau)$ [5.1, 5.2], i.e.:

$$K(\tau) = \overline{x(t)x(t+\tau)} \tag{5.5}$$

The line over the top of the product $x(t)x(t+\tau)$ indicates the time-averaging operation.

The correlation function is the measure of the size and duration of the after-effect of the fluctuations. For a very large number of practical cases the correlation function defines the full properties of the random process.

The spectral density, $S(f)$, and the correlation function, $K(\tau)$, of a stationary random process are related by a Fourier transformation pair. This is the Wiener-Khintchine theorem. For a stationary random process $x(t)$ the average $\overline{x(t)}$ and variance $\sigma_x^2 = \overline{x^2(t)} - \overline{[x(t)]}^2$ are constant quantities and they do not depend on time. The correlation function of such a process depends only on the difference, τ, of the two times considered. The power spectrum $S_x(f)$ is determined from the correlation function as [5.1, 5.2]:

$$S_x(\omega) = 2\int_{-\infty}^{\infty} K(\tau)e^{-i\omega\tau}d\tau = 4\int_{0}^{\infty} K(\tau)\cos\omega\tau d\tau \tag{5.6}$$

where $\omega = 2\pi f$ is the angular frequency.

The correlation function in turn is the Fourier transformation of the power spectrum $S_x(f)$.

$$K(\tau) = \frac{1}{4\pi}\int_{-\infty}^{\infty} S_x(\omega)e^{i\omega\tau}d\omega = \frac{1}{2\pi}\int_{0}^{\infty} S_x(\omega)\cos\omega\tau d\omega \tag{5.7}$$

This expression can be transformed to a form that is convenient for calculation [5.1, 5.3]:

$$K(\tau) = \int_{0}^{\infty} S_x(f)\cos 2\pi f\tau df \tag{5.8}$$

By evaluating this at $\tau = 0$ and with $\overline{x(t)} = 0$ we obtain the variance of the random variable $x(t)$:

$$\sigma_x^2 = \overline{x^2(t)} = K(0) = \int_{0}^{\infty} S_x(f)df \tag{5.9}$$

5.3   Electrical noise in solids

There are some types of electrical noise in solids that can be distinguished by their physical nature and mathematical description. In integrated circuits the basic sources of noise are the resistances, semiconductor diodes and transistors. The major types of noise are thermal, shot, generation–recombination (g–r) and flicker noise [5.1–5.3]. The generation–recombination (g–r) and flicker noise are also called excess noise and the flicker noise is also called, current noise, low frequency noise, or $1/f$ noise. Let us consider each type of noise separately. Although the shot and generation–recombination noises are not inherent for continuous metal films it is necessary to review them in order to understand the physics of electrical noise in solids.

5.3.1 Thermal noise

The thermal motion of the free electrons inside a conductor with non-zero resistance causes internal micro-currents with an average value equal to zero but the instantaneous values are non-zero. Because of the randomness of the thermal motion, at any time more electrons may move in one direction than in the other direction. Thus the instantaneous value of voltage between any points of a conductor will be a random function of time and small fluctuations of voltage will take place between the ends of the conductor with a magnitude depending on the resistance, R, and temperature, T, of the conductor. This is thermal (or kT or Nyquist or Johnson) noise.

If a capacitor, C, is connected across the resistor, the electrical charge on its plates will vary in absolute value and sign. The average value is equal to zero. The presence of the capacitor reduces the intensity of the noise at frequencies above $(RC)^{-1}$.

In their motion the electrons will collide with the crystal lattice and with imperfections of the lattice. These collisions contribute to the resistance and the amplitude of vibration of the lattice atoms is proportional to the temperature. It is possible to assume that the velocity of motion for the individual electrons inside a conductor remains constant between collisions and after a collision the electron changes its direction of motion and speed. At room temperature for metals the length of the mean free path of the conduction electrons is about 10^{-8} m. At room temperature the electrons move with rather large velocities (about $100\,\mathrm{km\,s^{-1}}$) and the time between impacts is about $\tau \sim 10^{-13}$ s, so that the elementary current impulses are of very small duration.

The total current is formed from a very large number of these elementary impulses of randomly positive and negative polarity. Therefore, on average, the number of electrons which pass through a cross-section of the conductor will be equal to zero. However at any given instant of time there may be an imbalance.

The PSD of thermal noise covers a broad frequency band. The upper boundary frequency of thermal noise at room temperature is $f_{\mathrm{max}} \approx 1/2\pi\tau$ $\sim 10^{12}$ Hz. At low frequencies the spectrum of the thermal noise is independent of frequency.

When a current flows in a metallic conductor under the operation of an electrical field, this directed component of the velocities will add to the random velocity of thermal motion of the free electrons. The velocity of the thermal motion electron is changed by only a very small amount, although this change is the source of the current in the conductor. In semiconductors the drift velocity may reach values that are comparable with the velocity of thermal motion. In both cases the directed increase in velocity and the random variations are independent so the thermal noise in metallic conductors does not depend on the magnitude of any direct current.

Let us now summarise the properties of the thermal noise:

1   The constant component of the noise current is equal to zero;
2   The instantaneous values of the current at each point on the time axis are distributed in a normal law;
3   The thermal noise level does not depend on the material of the resistor at the same sample temperature and resistance. The thermal noise is determined by the random motion of the current carriers and is a pure thermodynamic quantity.

Let us calculate a formula for the PSD of the thermal noise across a resistor. The thermal noise can be described by a voltage generator with mean square amplitude $[S_T(0)\Delta f]^{\frac{1}{2}}$ in series with the resistance as in Figure 5.1. This noise voltage generator is a source with a uniform spectrum from zero frequency up to very high frequencies. Such a generator is called a white noise generator.

Let us consider the RC-circuit shown in Figure 5.1 but feed it with a sinusoidal voltage $U = U_0 \exp(i\omega t)$ instead of the generator U_T. The voltage on the capacitor, U_C, will be:

$$U_C = \frac{U_0}{1 + i\omega RC} \quad \text{and} \quad U_C^2 = \frac{U_0^2}{1 + (\omega RC)^2} \tag{5.10}$$

Thus the contribution to the voltage on the capacitor from the noise generator in a frequency interval df is given by:

$$\overline{dU_C^2} = \frac{S_T(0)df}{1 + (\omega RC)^2} \tag{5.11}$$

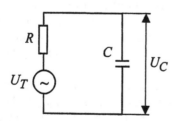

Figure 5.1 Electrical circuit for the derivation of the Nyquist formula. The thermal noise of the resistor is represented by the mean square EMF $U_T = \sqrt{S_T(0)\Delta f}$.

and the average intensity of the voltage $\overline{U_C^2}$ on the capacitor is:

$$\overline{U_C^2} = \int_0^\infty \overline{dU_C^2} = S_T(0) \int_0^\infty \frac{df}{1 + (\omega RC)^2} \tag{5.12}$$

After integration we obtain:

$$\overline{U_C^2} = S_T(0)/4RC \tag{5.13}$$

and for the thermal noise PSD we have:

$$S_T(0) = 4RC\overline{U_C^2} \tag{5.14}$$

According to the law of equipartition of energy on the degrees of freedom [5.2] for the circuit of Figure 5.1 with one degree of freedom we have:

$$\frac{1}{2}C\overline{U_C^2} = \frac{1}{2}kT \tag{5.15}$$

where k is the Boltzmann constant ($k = 1.38 \times 10^{-23} \, \mathrm{J\,K^{-1}}$) and T is the absolute temperature.

Hence we obtain the equation:

$$\overline{U_C^2} = \frac{kT}{C} \tag{5.16}$$

After substituting equation 5.16 into equation 5.14 we obtain an expression for the thermal noise PSD:

$$S_T(0) = S_T(f) = 4kTR \tag{5.17}$$

and the mean square thermal noise voltage for a conductor with resistance R is determined by the Nyquist formula:

$$\overline{U_T^2} = S_T(0)\Delta f = 4kTR\Delta f \tag{5.18}$$

For the any two-terminal impedance the more general Nyquist relation is:

$$\overline{U_T^2} = 4kT\mathrm{Re}Z(f)\Delta f \tag{5.18a}$$

where $Z(f)$ is the AC conductor impedance (complex resistance) in Ohms, Δf is the frequency band passed by the circuit or the measuring instrument, and Re means the real part. The value of $\overline{U_T^2}$ is expressed in units of V^2.

It is also possible to represent the Nyquist relation for the mean square current of thermal noise, $\overline{I_T^2}$ as:

$$\overline{I_T^2} = 4kT\mathrm{Re}Y(f)\Delta f \tag{5.18b}$$

where $Y(f)$ is the AC admittance (complex conductance) of the sample. The noise voltage and noise current are related through $\overline{U_T^2} = |Z(f)|^2 \overline{I_T^2}$.

Using the Nyquist relation, one may evaluate the thermal noise level in any two-terminal device with a non-linear current-voltage characteristic (CVC). In this case the differential, dynamic, resistance or differential conductance should be substituted in equations 5.18a or 5.18b instead of the values of $\mathrm{Re}Z(f)$ or $\mathrm{Re}Y(f)$.

Thermal noise is generated in any element of an IC that shows dissipation of energy if a current is passed through it. Pure reactive elements, which do not possess resistance, do not contribute to the thermal noise.

The Nyquist formula is not valid over all frequencies, but only for frequencies for which we can neglect quantum effects, i.e. for frequencies for which the relation $h/kT \ll 1$ is true and the Plank constant is $h = 5.62 \times 10^{-32}\,\mathrm{J\,s}$. In this case the quantum of the energy hf is much less than the thermal energy kT. If the value of hf/kT becomes comparable with unity, the quantum modification of the Nyquist formula is necessary. This is [5.2]:

$$S_T(f) = 4Rhf \left(\exp\frac{hf}{kT} - 1 \right)^{-1} \tag{5.19}$$

This equation is obtained from equation 5.17 if the average energy of the quantum oscillator $hf\,[\exp(hf/kT) - 1]^{-1}$ (neglecting the zero point energy) is substituted in it instead of the classical energy kT.

Quantum effects cause a decrease in the thermal noise intensity at frequencies $f > kT/h$. For $hf/kT \ll 1$ equation 5.19 becomes equation 5.17. At room temperature, $T_0 = 300\,\mathrm{K}$, the inequality $hf/kT_0 \ll 1$ holds even for millimetric waves since the characteristic frequency $f_0 = kT_0/h = 5.3 \times 10^{12}\,\mathrm{Hz}$, which corresponds to a wavelength of $\lambda = 5 \times 10^{-2}\,\mathrm{mm}$.

For a wavelength $\lambda = 3\,\mathrm{cm}$ corresponding to the frequency $f = 10^{10}\,\mathrm{Hz}$ the Nyquist equation 5.17 holds down to the temperature of liquid helium, $T = 4.2\,\mathrm{K}$, but it does not hold at helium temperatures for frequencies corresponding to wavelengths in the millimetric range. Therefore it is possible to use the classical limit of equation 5.17 at the usual operation temperatures of electron devices in all range of the radio frequency.

As the PSD of thermal noise is constant over a broad band of frequencies up to optical, the thermal noise is called "white" noise. At room temperature ($T_0 = 300\,\mathrm{K}$) the Nyquist equation 5.18 can be written:

$$\sqrt{\overline{U_T^2}} = U_T = 1.26 \times 10^{-4} \sqrt{R\Delta f} \tag{5.20}$$

where U_T is the value of the thermal noise effective voltage in units of μV, R is expressed in Ω and Δf is in Hz. For example, for a resistance $R = 10^4\,\Omega$, in a band width $\Delta f = 1000\,\mathrm{Hz}$ the voltage is $U_T = 0.4\,\mu V$.

The value $kT\Delta f$ is the maximum power of thermal noise that can be dissipated in an output circuit in the band Δf, when the impedances are matched. This is easily obtained from the circuit of Figure 5.2, where the resistance R is the source of thermal noise, which is fed to a noiseless load R_1. The thermal noise is represented by the mean square EMF $U_T = \sqrt{S_T(0)\Delta f}$. From Figure 5.2:

$$\overline{I_T^2} = \frac{\overline{U_T^2}}{(R + R_1)^2} \tag{5.21}$$

the power P going in to the load R_1 is equal to:

$$P = \overline{I_T^2}R_1 = \frac{\overline{U_T^2}R_1}{(R + R_1)^2} \tag{5.22}$$

The maximum power dissipated in a load R_1 is achieved when $R = R_1$, to give, after substitution of 5.18 in to equation 5.22.

$$P_{max} = \frac{\overline{U_T^2}}{4R} = kT\Delta f \tag{5.23}$$

Note that the Nyquist equation 5.18 is only applicable to systems in thermodynamic equilibrium. In such equilibrium systems the well-known Einstein relation:

$$\frac{D_0}{\mu_0} = \frac{kT}{e} \tag{5.24}$$

which connects the diffusion coefficient of the current carriers in a semiconductor, D_0, with the drift mobility, μ_0, holds, where e is the electronic charge.

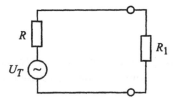

Figure 5.2 Electrical circuit for the evaluation of the formula for the thermal noise power transferred to a matched noiseless load R_1. The thermal noise of the resistor R is represented by the mean square EMF $U_T = \sqrt{S_T(0)\Delta f}$.

5.3.2   Hot electron noise

Let us apply a voltage to a semiconductor layer. At low bias, the current flowing through the device is proportional to the applied bias (Ohmic regime). But at higher voltages the current is no longer proportional to the voltage and the hot carrier regime is said to take place.

The electron gas in a semiconductor subjected to a strong electric field is no longer in equilibrium since the average energy of the electron motion increases and becomes more than the equilibrium thermal value, which is equal to $3/2(kT_0)$. Thus heating of the electrons occurs and these are called hot electrons. Normally an electric field is considered strong when the current through a homogeneous semiconductor sample no longer obeys Ohm's law. The physical mechanism of the non-linearity of the current–voltage characteristic in a strong electric field is the non-linear dependency of the electron drift mobility on the electric field.

Under non-equilibrium conditions the Nyquist formula does not hold and the Einstein relation, equation 5.24, is also not valid. In a non-equilibrium electron plasma of a semiconductor, as well as the thermal noise, there is an additional noise connected with the random nature of the loss of energy from the hot electrons and the lattice and with fluctuations of the dissipated power from an external source. This additional noise arises because of fluctuations of the average electron energy. This sort of noise is called diffusion, or hot electron noise and is a more general type of electrical noise than the thermal noise [5.2, 5.3]. These fluctuations give a mechanism for the electron energy dispersion and the thermal noise in such a physical system is that part of the fluctuations which is due to only the thermal energy.

If the Einstein relation (equation 5.24) holds, the diffusion noise reduces to thermal noise. The origin of the diffusion noise is connected to the dependence of the electron mobility on the average energy. The connection between the non-linearity of the current–voltage characteristic for a GaAs semiconductor sample and its diffusion noise power has been established experimentally.

To account for the noise in the hot carrier regime of semiconductors the concept of a noise temperature $T_n(f)$ has been introduced [5.2]. For a two-terminal device this is defined by one of the two equivalent equations:

$$S_U(f) = 4kT_n(f)\mathrm{Re}[Z_d(f)] \tag{5.25a}$$
$$S_I(f) = 4kT_n(f)\mathrm{Re}[Y_d(f)] \tag{5.25b}$$

where $Z_d(f)$ and $Y_d(f) = 1/Z_d(f)$ are respectively the AC, differential impedance and admittance at the DC bias point of the current–voltage characteristic of a semiconductor device.

It follows from this definition of the noise temperature that it is necessary to determine experimentally the spectral density of the voltage fluctuations S_U, or current fluctuations S_I, and the AC impedance Z_d, or AC admittance Y_d, of a semiconductor sample.

The concept of noise temperature is widely used in radio-physics to evaluate the noise level in those semiconductor devices intended for the amplification and transformation of electrical signals. The sensitivity of various semiconductor devices in the microwave range is evaluated using the noise temperature and is usually determined by comparison of the device noise with a standard noise generator.

In conditions of thermodynamic equilibrium the noise temperature coincides with the thermodynamic temperature of a system. In non-equilibrium conditions the noise temperature depends on the frequency of the noise measurement, the magnitude of the electric field creating the non-equilibrium conditions and on the details of the noise measurement. Although in non-equilibrium conditions the Einstein relation (equation 5.24) does not hold it is possible to show [5.3], that the following equation holds:

$$\frac{D}{\mu_d} = \frac{kT_n}{e} \tag{5.26}$$

where μ_d is the differential mobility of electrons at the operating point of the semiconductor device, $\mu_d = dv_0/dE$, where v_0 is the drift velocity. This formula is the analogy of the Einstein relation (equation 5.24) for the non-equilibrium case of the hot electron gas. With the help of equation 5.26 one can determine the noise temperature of the hot electrons from measurements of the diffusion coefficient and the differential mobility.

In modern microwave devices based on GaAs heterostructures with quantum wells, for example in high electron mobility transistors (HEMTs) the two-dimensional electronic gas is subjected to a high electric field, which is directed parallel to the plane of the electronic gas. In such devices the noise of the hot electrons may be dominant in the microwave range at low operating temperatures.

5.3.3   Shot noise

Shot noise in conductors is caused by the discreteness of the current flow produced when the individual electronic charges flow independently through a device. They act independently when they are excited up the potential step in a semiconductor diode, or when emitted from a hot cathode in a vacuum diode or from a photocathode.

The shot noise arises because of the discreteness of the charge. The term "shot" derives from the similarity of the electron current to small shot pellets. To describe the mechanism of the origin of shot noise it is more convenient to use the example of a vacuum diode. In a vacuum diode the electrons are emitted from a hot cathode and, under the operation of an electric field between the cathode and anode, pass across the cathode–anode gap and reach the anode. It is possible to consider the instants of the electrons' departure from the cathode and arrival at the anode as random independent events. If no space charge is formed in the diode, all the electrons leaving the cathode reach the anode. The diode is said to operate in the saturation regime. Thus each electron creates an impulse of anode current in the external circuit with a duration determined by the time of flight through the distance from the cathode to the anode. The resulting anode current, $I(t)$, formed from the separate electrons has fluctuations $\Delta I(t)$ around the average value, I_0, so that $I(t) = I_0 + \Delta I(t)$ (Figure 5.3). The average value of the noise current $\Delta I(t)$ is equal to zero. For a vacuum diode operating in the saturation regime the current fluctuation intensity about the average anode current is expressed by the Schottky formula [5.3]:

$$\overline{i_{sh}^2} = 2eI_0\Delta f \qquad (5.27)$$

with a spectral intensity:

$$S_{sh} = 2eI_0 \qquad (5.28)$$

If there is a space charge in the diode, the shot noise intensity is reduced and it is equal to:

$$S_{sh} = 2e\Gamma^2 I_0 \qquad (5.28a)$$

where Γ is the suppression factor.

The upper boundary frequency f_{max} of the shot noise PSD is defined by the time of flight of the electrons from cathode to anode, τ, and is given by $f_{max} \sim 1/2\pi\tau$. For the majority of vacuum diodes the time of flight is $\tau \sim 10^{-9}$ s. Therefore the PSD of shot noise appears uniform up to frequencies $f_{max} \sim 10^3$ MHz.

In semiconductor devices the shot noise is generated by the electrons passing across a potential step. The magnitude of the shot noise depends on the polarity and value of the voltage across the p–n junction. The shot noise in semiconductor devices is created not only by the majority carriers

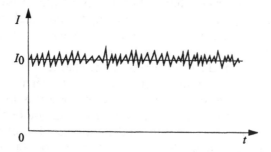

Figure 5.3 The fluctuations of the diode current in time determined by the discreteness of the electric charge.

but also by the minority carriers, whose contribution to the noise level depends on the operating regime of the device.

Let us consider the shot noise in a p–n junction [5.3]. The average current of an ideal p–n junction is described by the equation:

$$I = I_s \exp\left(\frac{U}{\varphi_T} - 1\right) \tag{5.29}$$

where I_s is the saturation current, U is the voltage on the p–n junction, $\varphi_T = kT/e$ is the thermal potential.

The total current consists of two independent currents flowing in opposite directions in the p–n junction: a diffusion current of the minority carriers $I_s \exp(U/\varphi_T) = I + I_s$ and a drift current of the majority carriers, I_s.

Each of these currents fluctuates independently and creates a shot noise. Thus by analogy with equation 5.28 the PSD of the low frequency shot noise in a semiconductor diode is [5.3]:

$$S_{sh} = \frac{\overline{i_{sh}^2}}{\Delta f} = 2e\,I_s \exp\frac{U}{\varphi_T} + 2e\,I_s = 2e(I + 2I_s) \tag{5.30}$$

The upper frequency of the shot noise spectrum is determined by the length of the independent impulses of current of the majority and minority carriers. The length of the impulses of the diffusion current is determined by the diffusion time of the minority carriers through the space charge. These carriers contribute to the full conductance of the p–n junction:

$$G = G_n + G_p = \frac{I + I_s}{\varphi_T} \tag{5.31}$$

where G_n and G_p are the conductances contributed by the electron and hole diffusion currents.

The drift current created by the majority carriers does not influence the total conductance for voltages $|U| > 3\varphi_T$, since the saturation current does not depend on the voltage.

For $I \gg I_s$ equation 5.30 for the shot noise becomes:

$$S_{sh}(f) = 2eI \tag{5.32}$$

The shot noise in a diode can be presented by an equivalent generator of a current source in parallel with the p–n junction and with the noise PSD given by equation 5.30.

At zero voltage and total current we obtain from equation 5.30 an expression for the shot noise:

$$\overline{i_{sh}^2} = 4eI_s\Delta f \tag{5.33}$$

Let us now calculate the thermal noise of a p–n junction at equilibrium, zero voltage, which is given by the Nyquist equation 5.18b, through the differential resistance, where $\operatorname{Re} Y = r_d^{-1}$:

$$\overline{i_T^2} = \frac{4kT\Delta f}{r_d} \tag{5.34}$$

According to equation 5.31, at $I = 0$ the differential resistance of the p–n junction is equal to:

$$r_d = \frac{1}{G} = \frac{\varphi_T}{I_s} = \frac{kT}{eI_s} \tag{5.35}$$

After a substitution of r_d in equation 5.34 we obtain:

$$\overline{i_T^2} = 4eI_s\Delta f \tag{5.36}$$

We see that at the equilibrium condition ($U = 0$) the shot noise level given by equation 5.33 coincides with the thermal noise level of the p–n junction. This result is the consequence of the fluctuation–dissipation theorem, which states that in thermodynamic equilibrium there is no noise other than the thermal noise.

In homogeneous conducting materials the shot noise is connected with the drift velocity of the current carriers, whereas the thermal noise is connected with the thermal motion of carriers and, therefore, with their thermal velocity. In metals the drift velocity of the current carriers is very small in comparison with the thermal velocity. Thus the level of the thermal noise considerably exceeds that of the shot noise, which is not observed in metals. Also the motion of the carriers is not independent.

Note that thermal and shot noise cannot be absolutely eliminated and they define the fundamental limit below which noise in electronic devices cannot be reduced. The origin of the thermal noise is connected to the thermal motion of carriers which cannot be removed completely. The origin of the shot noise is connected with the discreteness of an electrical charge. By increasing the coherence of the current pulses the shot noise can be reduced below the maximum value given in equation 5.28.

As well as these fundamental noise contributions, other types of noise are observed in real devices. These are "in excess" of these contributions. They depend on the details of the device construction and materials and are observed as resistance or emission fluctuations.

5.3.4 Generation–recombination noise

In semiconductors an excess noise, generation–recombination (g–r) noise [5.2, 5.3], is observed. This is created by fluctuations in the rates of the free carrier generation and recombination so that the number of free carriers in the conduction or valence bands fluctuates. As with $1/f$ noise this is a resistance fluctuation.

This type of noise arises in semiconductor devices because of the presence of impurity atoms or defects in the crystal lattice, which produce trap levels in the forbidden band. A carrier may be free to conduct or may be trapped and immobile.

If there is only one type of trap, the normalised resistance intensity fluctuations of the g–r noise is determined by the equation [5.2, 5.3]:

$$\frac{S_{gr}(f)}{R^2} = \frac{4}{N_0^2} \frac{\overline{\Delta N^2} \tau}{1 + \omega^2 \tau^2} \tag{5.37}$$

where R is the resistance of the semiconductor sample with a fluctuating number of carriers N, N_0 is the equilibrium number of carriers; $\Delta N = N - N_0$, and $\overline{\Delta N^2}$ is the variance; τ is the average carrier lifetime, which is usually in the range of 10^{-6}–10^{-3} s.

It is possible to assume that $\overline{\Delta N^2} = \beta N_0$, where β is a constant that depends on the statistics of the carriers in a sample. For many practical cases $\beta = 1$.

If there are M distinct generation–recombination levels distinguished by their depth in the forbidden band, the g–r spectra (equation 5.37) are superposed and the g–r noise is given by:

$$\frac{S_{gr}(f)}{R^2} = \sum_{i=1}^{M} \frac{A_i \tau_i}{1 + \omega^2 \tau_i^2} \tag{5.38}$$

where A_i and τ_i are the constants characterising the ith g–r process.

Figure 5.4 shows the dependence of the g–r noise PSD on frequency for the case of one type of trap in a semiconductor. This follows from equation 5.37,

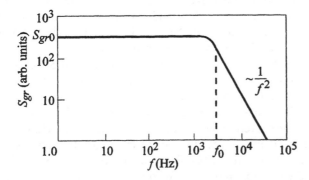

Figure 5.4 The power spectrum of generation–recombination noise.

and it can be seen in Figure 5.4, that at low frequencies up to some characteristic frequency the spectral intensity has a uniform spectrum where the g–r noise PSD is constant at S_{gr0}.

For frequencies where $\omega^2 \tau^2 \gg 1$ the g–r noise PSD, $S_{gr}(f) \sim 1/\omega^2$, so that it decreases with frequency with a square law. The spectrum of such a type is called a Lorentzian spectrum. From the characteristic frequency, f_0, at which $S_{gr}(f_0) = S_{gr0}/2$, it is possible to determine the average lifetime of the carriers $\tau = 1/2\pi f_0$.

In microwave devices based on GaAs, for example in field-effect transistors, the g–r noise originates from the interchange of the mobile charge carriers in the channel with deep donor levels and can be the largest source of noise. Note that at low measuring frequencies this noise may appear as an excess white noise.

5.3.5 1/f noise

In conducting films, and many other devices, there is a component of the noise with a power spectral density proportional to $1/f^\gamma$, where the exponent $\gamma \approx 1$ over a very wide range of frequency. This type of fluctuation is observed in practically all materials and components of circuits: in diodes, transistors, photo-resistors, bolometers, and also in many physical and biological systems. Noise of the $1/f^\gamma$ type is a very common type of fluctuation.

For different electron devices this type of noise has been observed in a broad band of frequencies from 10^{-6} to 10^6 Hz covering twelve and more decades [5.2]. However, in many modern devices 1/f noise is not observed at frequencies higher then several kilocycles at normal operating currents, since for these higher frequencies the 1/f noise is smaller than the white thermal noise that is always present.

This type of noise with a 1/f like spectrum is also referred to as excess or flicker (sometimes, current) noise. The term "1/f noise" usually implies that γ is close to unity. Flicker noise with a frequency exponent $0.8 \le \gamma \le 1.4$ is most commonly observed.

The magnitude of this noise at low frequencies can exceed the level of the thermal noise by tens and sometimes thousands of times. The term "excess noise" is also used for g–r and burst (see Section 5.3.5) noise. The 1/f noise limits the sensitivity and stability of many electronic devices.

The first studies of flicker noise were made by Johnson and Schottky in lamps with oxide cathodes 1925–1926. It was called "flicker noise" or the "flicker-effect" because of the slow fluctuations in the cathode emission. Since then there have been many investigations into flicker fluctuations in conducting materials. The current concepts of flicker noise in solids have been discussed and summarised in a large number of monographs and reviews [5.2–5.8]. Unlike the previously described fundamental types of noise, which are well understood, the physical nature of the origin of 1/f noise in all solids is not yet fully understood.

Excess fluctuations reflect many processes at the electron and atom level so that $1/f$ noise is an informative parameter for evaluating the quality of materials and the reliability of devices and ICs [5.7, 5.8].

Usually the PSD of flicker noise in Ohmic conducting films as a function of frequency f and current I can be approximated by the equation:

$$\frac{S_R(f)}{R^2} = K_1 f^{-\gamma} \qquad (5.39)$$

Here the constant K_1 and the frequency exponent γ are determined by the properties of the film material. The index γ is usually close to unity.

The $1/f$ noise PSD dependence shown in logarithmic co-ordinates, is a straight line and the slope gives the value of the exponent $\gamma = \Delta(\log S)/(\Delta \log f)$ as seen in Figure 5.5.

5.3.6 Burst (RTS) noise

In addition to the types of noise considered above, another type of electrical noise is observed in various solid-devices; it is called burst or random telegraph signal (RTS) noise [5.2, 5.3, 5.8]. Burst noise has been seen in many different devices including transistors, p–n junction diodes, Schottky diodes, different types of resistors, metallic point contacts and granular semiconductors. This phenomenon appears as a two-level random telegraph signal, that consists of a two-level process with transitions after random times, t_1 in the lower level and t_2 in the upper level with identical height of the current

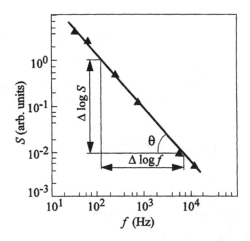

Figure 5.5 The power spectrum of $1/f$-like noise on a log–log scale. The slope represents the exponent and normally the magnitude at 1 Hz is quoted.

impulses (Figure 5.6). It is often found that a real device shows a more complex RTS pattern with several two-level or multilevel processes.

A signal with small amplitude, comparable with other noise contributions, is usually called RTS noise. A two-level signal occurs in some devices as a big signal involving large current pulses and this is called burst noise. It is supposed that the mechanism of burst noise in p–n junctions is connected with the irregular on-off switching of a channel at the surface or within the volume of the junction. The experimental data indicate that dislocations in the crystal structure are involved.

A random telegraph signal noise has also been observed in sub-micron devices, for example in MOS transistors [5.8]. The amplitude of these impulses is small on an absolute scale and corresponds to the transition of a single free carrier to a trap which produces changes in the resistance (conductivity) because of a decrease in the number of free carriers in the channel. In such field effect transistors it is possible to use this RTS noise to study the dynamics of the capture and emission of carriers into individual traps and to determine the physical parameters of these traps. The smaller the semiconductor structure the more likely one is to observe the operation of a single centre without the presence of other signals.

Burst noise has a Lorentzian spectrum similar to that of g–r noise [5.8]:

$$S_{\text{bur}}(f) = \frac{A\tau_0 I_0^2}{1 + \omega^2 \tau_0^2} \tag{5.40}$$

where A is a constant and τ_0 is the characteristic time defined by the relation:

$$\frac{1}{\tau_0} = \frac{1}{\tau_1} + \frac{1}{\tau_2} \tag{5.41}$$

where τ_1 and τ_2 are the average values of the times t_1 and t_2 in each state of the trap (Figure 5.6).

The development of sub-micron engineering promoted the study of RTS noise. Three types of excess noise producing fluctuations in the number carriers: g–r noise, flicker noise and burst noise have common features in

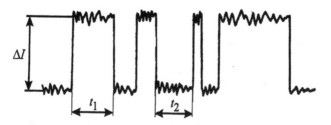

Figure 5.6 A typical bistable burst noise (two-level random telegraph signal) with white noise superimposed on it.

sub-micrometer size structures. With an increase in the device size the number of trapping centres increases. If the distribution of the number of centres with τ_0 is suitable then the RTS noise changes into noise with a $1/f$ like spectrum [5.8, 5.9]. The necessary distribution is one that is uniform in the exponent if τ_0 varies exponentially. This is often the case.

5.4 Physical mechanisms of $1/f$ noise in solids

5.4.1 Common requirements of the theory

The experimental fact that $1/f$ noise is observed over a wide frequency band indicates that physical processes with a wide distribution of relaxation times are involved. Thus it is necessary to find a physical mechanism for conduction fluctuations in solids which would give a broad distribution of relaxation times from a shortest time $\tau_{min} = 1/2\pi f_H$, where f_H is the upper measurable frequency of the spectrum, up to the longest time $\tau_{max} = 1/2\pi f_L$, where f_L is the lowest measurable frequency of the $1/f$ noise spectrum. For the lowest frequency $f_L = 10^{-6}\,\text{Hz}$ the longest time constant of this process that has been observed is of the order of magnitude: $\tau_{max} \approx 1.6 \times 10^5\,\text{s} \approx 4$ months. Since these measurement frequency limits are given by the superposition of the white noise at the high end and the measurement time at the low end the range will be larger for the process itself.

Various models have been suggested to account for the $1/f$ noise in solids, each usually designed to be used in a specific case. There are two principal approaches. One considers equilibrium models generating stationary noise, while the other is concerned with models that describe non-stationary processes which result in an approach to thermodynamic equilibrium or ageing. The former approach assumes that the $1/f^\gamma$ spectrum occurs in a limited frequency range between characteristic frequencies f_L and f_H. The objective is to provide a theoretical justification of the frequency range boundaries and the temperature dependence of the $1/f$ noise.

The $1/f$ fluctuations in homogeneous materials, resistive layers, semiconductors and metal films are due to conduction fluctuations which are known to occur even with no current flowing through the sample. It is possible to introduce many mechanisms of conduction fluctuations in homogeneous conductors or semiconductors: fluctuations in the number of carriers and their mobility, fluctuation of temperature, fluctuation of the geometric size of a sample, nucleation and annihilation of micro-defects in the crystal lattice.

If we assume that the fluctuations of voltage across an Ohmic sample at a given direct current arise due to fluctuations of the current-independent resistance, the PSD of the voltage fluctuations may be characterised by a quadratic dependence with current. Let us show this. Generally it is possible to represent fluctuations in the resistance of a homogeneous resistor or semiconductor layer as:

$$R(t) = R_0 + \Delta R(t) \tag{5.42}$$

where R_0 is the average value of the resistance; $\Delta R(t)$ is a fluctuation in the resistance which is a random function of time so that for a stationary process $\overline{\Delta R(t)} = 0$. In practice the level of fluctuations that is observed is very small, $\Delta R/R_0 \approx 10^{-10}-10^{-6}$. The largest fluctuations are observed in granulated materials, in particular in carbon resistors.

If a direct current is passed through a sample there are fluctuations of voltage between the ends $\Delta U(t) = I\Delta R(t)$. Thus the mean square of the noise voltage in a frequency band Δf at any frequency f is equal to:

$$\overline{\Delta U^2} = I^2 \overline{\Delta R^2} \tag{5.43}$$

The noise PSD is found from equation 5.43 using equation 5.3:

$$S_U(f) = \lim_{\Delta f \to 0} \frac{\overline{\Delta U^2}}{\Delta f} = I^2 S_R(f) \tag{5.44}$$

where $S_R(f)$ is the spectral density of the resistance fluctuations expressed in units of $\Omega^2 \mathrm{Hz}^{-1}$:

$$S_R(f) = \lim_{\Delta f \to 0} \frac{\overline{\Delta R^2}}{\Delta f} \tag{5.45}$$

The quadratic current law of the noise PSD is not always observed because sometimes the sample resistance is a non-linear function of the current. The resistance may be represented as a power series with coefficients fluctuating in time [5.10]:

$$R(t) = \sum_{k=0}^{n} R_k(t) I^k \tag{5.46}$$

The current-independent term, $R_0(t)$, defines the linear part of the current–voltage characteristic of the sample, Ohm's law. This term reflects the equilibrium fluctuations of resistance (conductance), which exist in the absence of any current flowing through the sample.

The coefficients of the higher order terms contribute to the resistance, and hence to its fluctuations, only if the current $I > 0$. Therefore, the fluctuations of these coefficients may be related to non-equilibrium (non-linear) conduction fluctuations which could be responsible for both the stationary and non-stationary noise (see Chapter 6). Non-equilibrium fluctuations cause deviations from the quadratic dependence in equation 5.44 and the magnitude of the fluctuations in equations 5.42 and 5.45 will depend on the current. This may be due either to different mechanisms of scattering of the current carriers or to the local self-heating of the film by Joule dissipation. For small currents, only equilibrium $1/f$ noise is observed.

It has been well established in experiments on non-metal samples that $1/f$ noise often arises from fluctuations of the linear part of resistance. Voss and Clarke [5.11] obtained direct experimental evidence of equilibrium resistance fluctuations by measuring the PSD of fluctuations with a $1/f$ spectrum in the thermal noise of InSb films and niobium island films in the absence of any current flowing through the samples. These fluctuations have also been observed when a direct, sinusoidal or pulse current is passed through a sample [5.12]. All these measuring techniques revealed identical $1/f$ spectra. These findings indicate that neither direct nor alternating current is necessary for $1/f$ noise, which is actually an equilibrium resistance fluctuation.

The main mechanism of conductivity fluctuations in semiconductors or metals depends on many factors. The $1/f$ noise can occur in semiconductors and metals due to fluctuations in the number or mobility of carriers since the resistance depends on both of these quantities. However, for metals the $1/f$ noise is due to mobility fluctuations since the carrier concentrations are so large and there are no significant trapping mechanisms.

5.4.2 Distribution of relaxation-times model

The model of an exponentially wide relaxation-time distribution [5.2, 5.5, 5.7] is well established for metal films. It is based on the summation of relaxation processes with different relaxation times. It considers $1/f$ noise as a superposition of random relaxation processes characterized by a relaxation time distribution which remains continuous between τ_{max} and τ_{min} ($\tau_{max} \gg \tau_{min}$) and is described by the distribution function $g(\tau)$.

In the simplest case, when the fluctuation kinetics of a random quantity $x(t)$ is characterised by a single relaxation time the PSD has a Lorentzian spectrum (5.37):

$$S_x(f) \propto \frac{\tau}{1 + \omega^2 \tau^2} \tag{5.47}$$

If the fluctuations of a random quantity $z(t)$ are defined by a set of relaxation times, with a continuous distribution function $g(\tau)$, the PSD of $z(f)$ is found by integrating equation 5.47 with the statistical weight $g(\tau)\mathrm{d}\tau$.

$$S_z(f) = \int_0^\infty S_x(f)g(\tau)\,\mathrm{d}\tau \tag{5.48}$$

Specifically, if the weighting function obeys the statistical law $g(\tau) \propto 1/\tau$ ($\tau_{min} \leq \tau \leq \tau_{max}$) then equation 5.48 gives the $1/f$ spectrum in the frequency range $f_L \ll f \ll f_H$ where $f_L = 1/2\pi\tau_{max}$ and $f_H = 1/2\pi\tau_{min}$.

The desired distribution $g(\tau)$ is easily obtained if the $1/f$ noise is due to an exponential process with a uniform distribution in the exponent. This also gives the large span of the frequency for a small span of the variable in the exponent. Such processes are activation processes in solids, generation–recombination and tunneling processes in semiconductors, the creation and annihilation of quasi-equilibrium vacancies in metals (see Sections 5.4.3 and 5.5.2).

It was shown in Section 1.3.3 that the mean time of the atom in a potential well of depth E_a is $\tau = \tau_0 \exp(E_a/kT)$, where τ_0 is the atomic thermal vibration period ($\tau_0 \sim 10^{-13}$ s).

When the activation energy, E_a, is uniformly distributed over a sufficiently broad interval, i.e. $G(E_a) = \text{const.}$ ($E_1 \leq E \leq E_2$), the following equation holds for the $1/f$ spectrum boundaries:

$$f_L = f_0 \exp\left(-\frac{E_2}{kT}\right) \text{ and } f_H = f_0 \exp\left(-\frac{E_1}{kT}\right) \qquad (5.49)$$

Here, $2\pi f_0 = \tau_0^{-1}$ is the average frequency of atomic thermal vibrations.

For an activation energy distribution from E_1 to E_2, the number of frequency decades with a $1/f$ spectrum is derived from equation 5.49 and can be very large:

$$f_H/f_L = \exp\left(\frac{E_2 - E_1}{kT}\right) \qquad (5.50)$$

For the $1/f$ law to be strictly fulfilled, the energy distribution function must be continuous. However, since noise data has inherent uncertainty, the $1/f$ spectrum can be simulated with reasonable accuracy by a very small discrete set of processes e.g., when there is only one relaxation process per frequency decade [5.13].

We will use the model of an exponentially wide relaxation-time distribution to explain the $1/f$ noise in semiconductors using a number fluctuation model, and in metal films, using a vacancy model, in Sections 5.4.3 and 5.5.2.

5.4.3   Carrier number fluctuation (Δn) model

In semiconductors, there are fluctuations of the free carriers number, Δn, because of g–r processes or the capture and emission of carriers at traps which can produce conductivity fluctuations. If the relaxation process is characterized by one trap level with a time constant τ, the noise has a Lorentzian spectrum and it is described by equation 5.37 for a generation–recombination noise.

If there are many independent centres in the semiconductor, with a wide distribution of relaxation times, $g(\tau)$, determined by the depth of the trap levels, with statistical weight [5.13]:

$$g(\tau)d\tau = \begin{cases} \dfrac{d\tau/\tau}{\ln(\tau_{max}/\tau_{min})}, & \text{for } \tau_{min} \leq \tau \leq \tau_{max} \\ 0, & \text{for } \tau < \tau_{min}, \tau > \tau_{max} \end{cases} \qquad (5.51)$$

the summation of the g–r spectra (equation 5.37) in accordance with equation 5.48 over all values of τ, and assuming that $\overline{\Delta N^2} = \beta N_0$, gives:

$$
\begin{aligned}
\frac{S_R(f)}{R^2} &= \frac{4\beta}{N_0} \int_{\tau_{\min}}^{\tau_{\max}} \frac{1}{1 + \omega^2 \tau^2} \frac{d\tau/\tau}{\ln(\tau_{\max}/\tau_{\min})} \\
&= \frac{4\beta}{\omega N_0 \ln(\tau_{\max}/\tau_{\min})} (\arctan \omega\tau_{\max} - \arctan \omega\tau_{\min})
\end{aligned}
\tag{5.52}
$$

For $1/\tau_{\max} \ll \omega \ll 1/\tau_{\min}$ the following approximate dependency of the $1/f$ noise PSD on frequency is obtained from equation (5.52):

$$
\frac{S_R(f)}{R^2} \approx \frac{\beta}{f N_0 \ln(\tau_{\max}/\tau_{\min})} \sim \frac{1}{f}
\tag{5.53}
$$

Since, if $\omega\tau_{\max} \gg 1$ the function arctan $\omega\tau_{\max} \approx \pi/2$ and if $\omega\tau_{\min} \ll 1$ the magnitude of the function arctan $\omega\tau_{\min} \approx 0$. The superposition of relaxation processes produces a power spectrum as $1/f$ in the frequency range $f_L \ll f \ll f_H$. From equation 5.53 it is seen that the noise PSD is inversely proportional to the number of carriers, N_0, in the sample.

The long relaxation times, $\tau_{\max}$, are usually associated with the quantum mechanical tunnelling of carriers to traps located inside the gate oxide of the MOS system. For MOS transistors the McWhorter model of the electronic interaction between a dielectric and the channel in a semiconductor explains the existence of time constants distributed in the broad band from 10^{-5} s to 10^8 s. The $1/f$-like noise results from fluctuations of the number of carriers in the channel because of the random capture and emission by the traps located in the oxide near the semiconductor–dielectric boundary. The relaxation times are determined by the time it takes for carriers to tunnel into traps in oxide. This varies exponentially with the distance between the trap position in the oxide and the semiconductor–dielectric boundary [5.2].

Thus, the low-frequency excess noise in semiconductors arises because of defects of the crystal. In this connection the $1/f$ noise may be used as a general measure of perfection and an indicator of quality of semiconductor devices during production to predict their reliability.

Although different distributions of relaxation-times models have successfully explained the $1/f$ noise in most systems, this has not been possible for many semiconductor devices. This has resulted in a theoretical model to explain why $1/f$ noise arises in semiconductors with a perfect lattice.

5.4.4 Mobility fluctuation ($\Delta\mu$) model

In 1969, Hooge and Hoppenbrouwers reported the observation of $1/f$ noise in continuous gold films. The noise magnitude was well described by the empirical formula which Hooge had previously derived for other conductors.

It describes the voltage or resistance fluctuations in a sample at constant current by:

$$\frac{S_U(f)}{U^2} = \frac{S_R(f)}{R^2} = \frac{\alpha_H}{N_0 f} \qquad (5.54)$$

Here, $S_U(f)$ and $S_R(f)$ are the spectral densities of the voltage and resistance fluctuations in the sample; N_0 is the number of carriers in the sample ($N_0 = V n_0$) where V is the sample volume and n_0 is the free carrier concentration. The dimensionless coefficient $\alpha_H = 2 \times 10^{-3}$ is referred to as the Hooge constant and equation 5.54 is called the Hooge formula. Note that it is not always easy to distinguish experimentally between the V^{-1} and N_0^{-1} dependencies. Further studies have demonstrated that the coefficient is not a single constant for all different conductors but should rather be used as a parameter to indicate the noisiness of the sample [5.15]. Then the relative noise intensity is defined by the relation:

$$\frac{S_U(f)}{U^2} = \frac{S_R(f)}{R^2} = \frac{\alpha}{N_0 f} \qquad (5.54a)$$

The parameter α has been measured experimentally for many semiconductor materials and metals. It is in the range $10^{-8} < \alpha < 10^{-1}$.

The hypothesis associated with the $1/f$ noise described by the Hooge formula 5.54a is that the noise is due to fluctuations of the carrier mobility and then only that part of the mobility which is due to the charge carrier scattering by phonons. The remaining part of the mobility, associated with scattering by impurities and lattice defects does not fluctuate. This hypothesis is supported by experiments in which the noise decreases with a decrease in the relative contribution of scattering by phonons to the overall resistance of p- and n-type silicon samples [5.15]. The $1/f$ noise was shown to decrease in strongly alloyed semiconductors and increase with temperature. The atoms of an alloying admixture in a crystal may be regarded as stable defects. The parameter α was also found to decrease as the carrier scattering from the boundaries of a bismuth film increased, as its thickness is reduced.

This hypothesis, which assumes that $1/f$ noise depends only on the scattering by phonons, gives rise to a modified equation for the Hooge parameter α in the form [5.15, 5.16]:

$$\alpha = \alpha_H (\mu/\mu_{lat})^2 \qquad (5.55)$$

where μ is the effective mobility of current carriers, and μ_{lat} is the lattice mobility.

Generally, several scattering mechanisms take place in a semiconductor or metal: (1) lattice (phonon) scattering; (2) impurity and lattice defect scattering. Each scattering mechanism on its own would lead to a value of the mobility. If both mechanisms are present simultaneously with a small concentration of impurities and lattice defects, the effective mobility of current carriers, μ, is given by:

$$\frac{1}{\mu} = \frac{1}{\mu_{\text{lat}}} + \frac{1}{\mu_{\text{df}}} \tag{5.56}$$

where μ_{df} is the value of the mobility with impurity and defect scattering only. This is Matthiessen's rule for conductors.

The essential part of the model is that it is only the μ_{lat} that fluctuates. For the PSD of the lattice and effective mobility fluctuations the following relations are true [5.15]:

$$\frac{S_\mu}{\mu^2} = \frac{\alpha}{fN_0} \quad \text{and} \quad \frac{S_{\text{lat}}}{\mu_{\text{lat}}^2} = \frac{\alpha_{\text{lat}}}{fN_0} \tag{5.57}$$

From equations 5.56 and 5.57, equation 5.55 for the Hooge parameter α then follows and the Hooge constant $\alpha_{\text{H}} = \alpha_{\text{lat}} = 2 \times 10^{-3}$ is connected with scattering by phonons. The model has been extended to a variety of inhomogeneous semiconductor devices [5.17].

It is assumed that the $1/f$ noise in the lattice scattering happens because of fluctuations in the scattering of the acoustic phonons. Experiments carried out on the scattering of light by acoustic phonons in quartz have shown that the intensity of the scattered light fluctuates with a $1/f$ spectrum [5.18]. There is still no analytical expression for the phonon noise. It is considered that the number of acoustic phonons in a mode fluctuates about an average value with a $1/f$ spectrum.

Equation 5.54a is correct if the experimentally determined PSD of the voltage fluctuations across the sample is directly proportional to the square of the applied voltage or current and inversely proportional to the frequency. Verification of the $S_{\text{U}} \propto U^2$ law makes it possible to use the relative fluctuation spectrum S_{U}/U^2 and interpret the $1/f$ noise as a result of equilibrium resistance fluctuations. When the quadratic dependence of S_{U} is not satisfied exactly, then the equation:

$$S_{\text{U}}(f) = \frac{\alpha U^{2+\theta}}{N_0 f^\gamma} \tag{5.58}$$

is also called the Hooge formula. Here, the parameter α is a dimensional quantity. The frequency exponent, γ, may differ from $\gamma = 1.0$ while the exponent θ defines the non-linearity of the current–voltage characteristic (see Chapter 6).

For the quadratic dependence of the noise PSD on the applied voltage and γ values close to unity, the noise level is often expressed by the dimensionless parameter α defined from equation 5.54a as:

$$\alpha = \frac{S_U(f)N_0 f}{U^2} \tag{5.59}$$

5.5   Noise in metal films due to structural micro-defects

5.5.1   Basic observations

Eberhard and Horn [5.19] suggested that the $1/f$ noise in metal films is due to vacancy diffusion, and the increase in its intensity with temperature stems from the increased concentration of vacancies. Robinson [5.20] hypothesized that $1/f$ noise in metals may originate from the random motion of defects in the lattice.

Pelz and Clarke [5.21] obtained experimental evidence for the relationship between the magnitude of the $1/f$ noise and the concentration of defects in polycrystalline copper films. The defects were induced by bombarding the films with fast 500 keV electrons while they were maintained at 90 K. It turned out that the change in film resistivity was proportional to the total number of defects. At the same time, the increase in the PSD of the $1/f$ noise expressed through the parameter α in accordance with equation 5.59 was found to vary as $\Delta\alpha \propto n_d^{0.6} \propto \Delta\rho_f^{0.6}$ as shown in Figure 5.7, where n_d includes

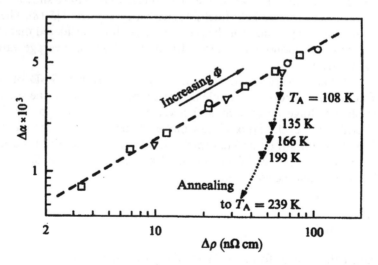

Figure 5.7 The changes in the intensity of the $1/f$ noise, expressed as the α parameter, against changes in the sample resistivity in Cu films with increasing radiation dose Φ [5.21]. The dashed line is drawn to show the dependence $\Delta\alpha \propto \Delta\rho^{0.6}$, while the solid triangles correspond to annealing at successively higher temperatures, T_A. For $T_A = 239$ K, $\Delta\rho \approx 11.6$ nΩ cm, and $\Delta\alpha \approx 7 \times 10^{-5}$. The different open symbols correspond to different samples.

the defects "frozen in" at the irradiation temperature. The annealing process reduced both the resistance and the noise.

The authors explained these findings by the generation of excess flicker noise only by mobile defects, which make up a small fraction of the total number. This explanation was verified in annealing experiments. In these the film is kept in a vacuum or inert gas at an elevated temperature for a certain time for the defects to be removed from the crystal lattice (see Section 1.5.3) and the defect concentration to reach equilibrium at that temperature. Mobile defects are readily annealed even at low temperatures (between 100 and 200 K), and this leads to a decrease in the excess noise by as much as an order of magnitude for a temperature rise from 100 to 200 K (Figure 5.7). The sample resistance in this temperature range does not vary with temperature since it largely depends on other types of induced defects, which anneal at higher temperatures (between 200 and 300 K). The additional resistance decreases by almost one order of magnitude for a rise in the annealing temperature from 200 to 300 K, while the excess noise remains practically unaltered over this temperature range. The annealing of these static defects also leads to a decrease in γ.

Pelz and Clarke [5.22] demonstrated that the PSD of the excess noise in Cu films alloyed with In is proportional to the additional resistivity caused by radiation damage. Also, bombardment with krypton, which creates more cluster defects than point defects, causes a smaller increase in the noise per unit increase in resistivity than electron irradiation. Vacancies are found to be the most "noisy" defects in a crystal.

5.5.2 The vacancy model of $1/f$ noise

Vacancies are thought to be the most important defects in metals because they require only a relatively small energy for formation and migration. Vacancies are involved in both the movement and the rearrangement of other defects. The contribution of vacancies to the resistivity is given by [5.23]:

$$\rho_v(t) = An_v(t) = A\frac{N_v(t)}{N_a} \qquad (5.60)$$

where A is the coefficient of proportionality, $n_v(t)$ is the atomic concentration of vacancies that fluctuates in time, and $N_v(t)$ and N_a are the numbers of vacancies and atoms in the sample, respectively. The concentration of equilibrium vacancies n_v is determined by equation 4.5.

Let us assume that the $1/f$ noise in metal films is produced by resistance fluctuations due to fluctuations in the number of vacancies in the sample with lifetime τ_v, which is a random variable [5.7]. The creation of vacancies increases the resistance during their lifetime.

The lifetime of vacancies is determined by the average distance between their source and their sink. The average number of vacancy jumps, η, between lattice sites from the times of creation to annihilation can be very large. In one study [5.24], the number of jumps for equilibrium vacancies in aluminium amounted to $\eta = 8 \times 10^7$. The total resistance change in a film due to vacancy number fluctuations can be described by a superposition of rectangular pulses of duration, τ_v. In this case, the relative voltage fluctuation spectrum across a sample with resistivity ρ_f is calculated using equation 5.60 through the spectrum of vacancy number fluctuations, S_{Nv}:

$$\frac{S_U(f)}{U^2} = \frac{S_{\rho_f}(f)}{\rho_f^2} = \frac{A^2}{\rho_f^2 N_a^2} S_{N_v}(f) \tag{5.61}$$

where $S_U(f)$ is the PSD of voltage fluctuations, $S_{\rho_f}(f)$ is the PSD of resistivity fluctuations, and $S_{N_v}(f)$ is the PSD of vacancy number fluctuations in the film.

Vacancy sinks in a homogeneous sample, such as a bulk metal or a film with a perfect structure, are uniformly distributed throughout the bulk. The probability of annihilation of each vacancy in each time interval during its lifetime is equal. The vacancy creation and annihilation events are statistically independent while the average lifetime of each vacancy equals [5.24]:

$$\tau_{v0} = \tau_0 \exp(E_v/kT) \tag{5.62}$$

where E_v is the activation energy of vacancy formation (see Section 4.2.2).

In this case, the power spectrum of the noise arising when a current I_0 flows through a sample containing N_v vacancies will have a Lorentz form [5.24]:

$$S_U(f) = 4\overline{\Delta R^2} I_0^2 N_v \frac{\tau_{v0}}{1 + \omega^2 \tau_{v0}^2} \tag{5.63}$$

where $\overline{\Delta R^2}$ is the variance of the resistance fluctuations in the specimen.

For the relative voltage fluctuation spectrum across a sample containing N_v vacancies one can write:

$$S(f) = \frac{S_U(f)}{U^2} = \frac{S_{N_v}(f)}{N_v^2} = 4N_v \frac{\tau_{v0}}{1 + \omega^2 \tau_{v0}^2} \tag{5.63a}$$

A voltage fluctuation spectrum with the form of equation 5.63, and associated with thermally activated vacancies, was observed by Celasco *et al.* [5.24] in experiments on Al films with a homogeneous microstructure and with a mean grain size of about 0.5 μm. Figure 5.8 shows the current noise power spectrum of Al films at two temperatures. As well as the noise induced by equilibrium vacancies, there is a low-frequency component with a $1/f^\gamma$-like spectrum, where $\gamma > 2$. The measurements were made at high temperatures where the non-equilibrium $1/f^\gamma$ noise is supposed to be due to atomic diffusion along the grain boundaries (see Section 7.3).

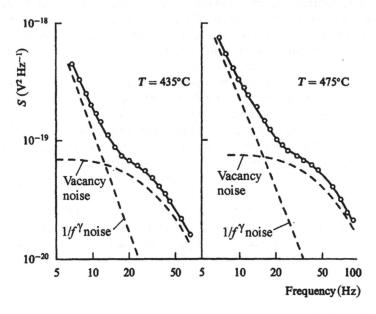

Figure 5.8 The energy spectra of current noise in thin Al films taken at two different temperatures with a current density $j = 3.6 \times 10^5 \, \text{A cm}^{-2}$ [5.24]. The dashed curves represent a best fit of the experimental data into components of equilibrium vacancy noise with a g–r spectrum and $1/f^\gamma$ noise ($\gamma > 2$). Points are averages over ten samples.

Thin metal films have an inhomogeneous structure. There are different vacancy sources (and sinks) distributed non-uniformly through the film volume. At sources and sinks vacancies may be created and annihilated. The internal energy of vacancy creation varies from one source to another because an atom may give rise to a vacancy by breaking a different number of bonds, depending on its position at the boundary of a grain, dislocation or step or at a void surface. Major vacancy sources and sinks in metals are grain boundaries, voids, and dislocations. It has been confirmed by experiment that, in films, vacancies are largely created at grain boundaries [5.24] since the mean vacancy diffusion range in Al films is comparable with the mean grain size.

The crystal structure of a real film is not in thermodynamic equilibrium and its free energy is increased due to the numerous defects. Because they are sources and sinks of vacancies, these defects change their shape and position upon emitting or absorbing a vacancy, for example, dislocations can move and voids grow. The concentration of micro-defects in a film is reduced by annealing. After annealing at low temperatures the defects are fixed, so that they move very slowly or do not move at all, and each source is characterised by a quasi-equilibrium concentration of vacancies. The emission or absorption of vacancies by a source does not lead to a change in the free energy of the crystal which thus remains in a quasi-equilibrium (local equilibrium) state, while the resulting flicker noise may be regarded as quasi-stationary and due to vacancy number fluctuations.

Numerous experiments have confirmed that external conditions, such as temperature or current, may cause a film to change to a non-equilibrium state and give rise to non-stationary $1/f$ fluctuations.

The creation of a vacancy in a perfect crystal requires that about half the atom's bonds in the lattice be broken, if we assume that the atom which leaves the bulk is moved to the surface of the crystal or a void. The energy of vacancy creation in metals is $u_v = 0.7$–$2.5\,\text{eV}$ [5.25], and the energy per atomic bond in the lattice equals $u_{v1} = 0.1$–$0.5\,\text{eV}$.

In thermodynamic equilibrium, the vacancy creation and annihilation rates are identical and the average vacancy concentration remains constant with time. The lifetime of vacancies depends on the distance, L_v, between their sinks and is defined by the relation [5.25]:

$$\tau_v = L_v^2 / \pi^2 D_v \tag{5.64}$$

where D_v is the vacancy diffusion coefficient.

The distribution of the lifetimes and hence the time constants is related to the distributions of the activation energy for vacancy diffusion and the distance between the vacancy sources and sinks, which are random variables for the film volume. The randomness of the activation energy is due to the variety of structural imperfections. This spread of time constants is responsible for the existence of $1/f$ noise over a wide frequency range.

Let us estimate the relaxation-time distribution limits at $T = 350\,\text{K}$. The minimum relaxation time corresponds to the vacancy lifetime (equation 5.62) for an atom to overcome a potential barrier equal to the energy needed to break a single bond in the crystal lattice. The assumption that $E_v = 0.1\,\text{eV}$ gives $\tau_{min} \approx 10^{-11}\,\text{s}$; so that for the highest frequency of the $1/f$ spectrum one finds $f_H \approx 1/2\pi\tau_{min} \approx 10^{10}\,\text{Hz}$.

The lower boundary frequency of the $1/f$ spectrum may be related to the vacancy lifetime (equation 5.64). A vacancy created at the boundary of a grain or a micro-void can diffuse through the bulk of a crystallite until it reaches a sink. If the distance between the source and sink is taken as $L_v = 20\,\text{nm}$ and the lattice diffusion coefficient for Al at $T = 350\,\text{K}$ is $D_v \approx 10^{-24}\,\text{m}^2\,\text{s}^{-1}$ [5.25], then, from equation 5.64, $\tau_{max} = 4 \times 10^7\,\text{s}$, so that $f_L \approx 1/2\pi\tau_{max} \approx 4 \times 10^{-9}\,\text{Hz}$. These estimates give evidence for a broad range of relaxation times for equilibrium vacancies.

If we assume that the variance of the vacancy number fluctuations in a sample is identical to their mean number, i.e. $\overline{\Delta N_v^2} = N_v$, and the distribution of relaxation times, $g(\tau)$, is given by equation 5.51, the PSD of vacancy number fluctuations in the frequency range from f_L to f_H can be written, after calculation of the integral (using equation 5.48) and taking into account equation 5.63a, as:

$$S_{N_V}(f) = \frac{N_v}{f \ln(f_H/f_L)}.$$ (5.65)

Substitution of equation 5.65 into equation 5.61 yields the relative PSD of voltage fluctuations in a film:

$$\frac{S_U(f)}{U^2} = \frac{A^2 n_v}{\rho_f^2 N_a f \ln(f_H/f_L)}$$ (5.66)

According to equation 4.5 for $n_v = \exp[-(u_v - \sigma V_v)/kT]$ and equation 5.66, the PSD of $1/f$ noise induced by fluctuations in the number of vacancies shows an activated temperature dependence and an exponential dependence on the mechanical stresses.

It follows from equation 5.66 that the parameter α, defined in equation 5.59 is

$$\alpha = \frac{A^2 N_0 n_v}{\rho_f^2 N_a \ln(f_H/f_L)}$$ (5.67)

5.5.3   Experimental results supporting the vacancy mechanism of $1/f$ noise

Dependence of the $1/f$ noise on film thickness

It is difficult to make experimental measurements of the dependence of the PSD of the $1/f$ noise on the film thickness because the magnitude of the noise varies in films of the same thickness due to differences in the structure and the concentration of micro-defects. The level of $1/f$ noise in a film may be tens of times higher or lower than in another film of identical thickness made using the same technology. The difference may exceed the effect being examined [5.6]. Films for these experiments must have a thickness $h > 10$ nm, since thinner layers are normally discontinuous with an island structure (see Section 1.5.4). Films with thickness $h > 1$ μm are inconvenient for experiment because it is difficult to pass a high enough current density in order to detect the flicker noise [5.5].

In order to study the thickness dependence of the noise, one must have films of different thickness with similar concentrations of micro-defects and impurities. Such films must be deposited under exactly the same conditions, for example by moving a shutter over the substrate during the condensation or by reducing the film thickness by anodising [5.7].

Numerous studies with metal films have shown that the $1/f$ noise arises in the bulk of the film, because the relative PSD of the fluctuations is inversely proportional to the volume, the thickness or to the number of carriers (or atoms) in the sample [5.5, 5.11, 5.15]:

$$\frac{S_U(f)}{U^2} \propto \frac{1}{V} \propto \frac{1}{h} \propto \frac{1}{N_0} \tag{5.68}$$

The Hooge formula (equation 5.54) is consistent with this dependence, which was reported to occur in thin gold [5.15] and platinum [5.26] films. The thickness of the platinum films was increased eight-fold ($h = 8$–$65\,\text{nm}$), while the number of atoms varied from 10^8 to 10^{14}.

If the appearance of $1/f$ noise is associated with the sample surface, for example, in the model of Celasco *et al.* [5.27], then the thickness dependence of the relative noise PSD is:

$$\frac{S_U(f)}{U^2} \propto \frac{1}{h^2} \tag{5.69}$$

According to this model, $1/f$ noise is generated in a thin subsurface layer near the film-substrate interface, while the rest of the film is "silent" and only shunts the subsurface source of the noise. This hypothesis is at variance with many experimental findings in metals.

The surface noise source observed in some experiments may have a different nature [5.7]. For example, it may be related to the drift of alkali metal ions in films deposited on glass or the effect of a gas layer absorbed by the substrate. These gases subsequently dissolve in the surface layer of the condensate to give rise to a contaminating atom-vacancy complex. This, in turn, leads to a high vacancy concentration in the film and an elevated noise level. The formation of such complexes was confirmed by the measurement of the internal mechanical stresses in molybdenum films.

A surface noise source can also arise from the non-uniform distribution of mechanical stresses across the thickness of the film. The stresses are thought to be especially strong at the film-substrate interface. Under certain conditions, the stresses in the subsurface layer can make a substantial contribution to the $1/f$ noise of the film.

The thickness dependence of the PSD may have quite a different form from that given by formulas 5.68 and 5.69. Figure 5.9 shows such a dependence for films on a glass substrate [5.7]. The film thickness was altered by the movement of a shutter. The figure also shows the thickness dependence of the void density, n_p (right scale). The lowest noise and the highest macro-void density were observed in the thickness range between 80 and 100 nm. The minimum resistivity also occurred within this range.

The observed effects can be accounted for by the thermodynamic advantage of moving a vacancy from the bulk of the film to macro-voids at a given thickness (see Section 4.3). For this reason, films of thickness $h \approx 100\,\text{nm}$ have the lowest concentration of vacancies. The rapid rise in noise intensity

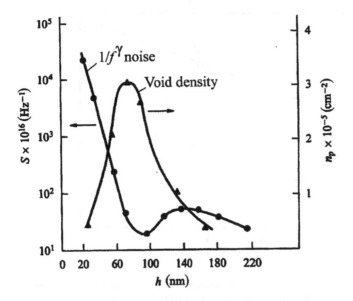

Figure 5.9 The thickness dependence of the PSD of $1/f$ noise at 20 Hz and the density of voids through the film for Cr films on glass [5.7]. The condensation rate was $w_c = 0.5$ nm s^{-1} and the condensation temperature $T_c = 470$ K.

with decreasing thickness (at $h < 40$ nm) may be due to the structural inhomogeneity of thin films. In films thicker than 150 nm, macro-void formation is thermodynamically disadvantageous. The generation of $1/f$ noise in such films is a volume effect related to fluctuations of the vacancy number in the sample even though the dependence of the noise PSD on the film thickness (see Figure 5.9) does not follow equation 5.68.

The vacancy diffusion origin of macro-voids is confirmed by the dependence of the void density on the mechanical stresses given by equation 4.20. In Figure 5.10 the experimental dependence of the macro-void density on mechanical stress for aluminum and chromium films are given. The dependencies are well approximated by the expression:

$$n_p \sim \exp(\sigma V_{ap}/kT) \tag{5.70}$$

The activation volume, V_{ap}, can be derived from Figure 5.10 using the equation:

$$V_{ap} = 2.3 \, kT \frac{\Delta \log n_p}{\Delta \sigma} \tag{5.71}$$

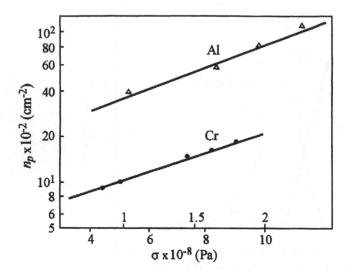

Figure 5.10 Dependence of the density of macro-voids on the mechanical stresses in aluminium and chromium films.

to give $V_{ap} = 0.6 \times 10^{-29}\,m^3$ for chromium films and $V_{ap} = 3.7 \times 10^{-29}\,m^3$ for aluminium. These magnitudes are close to the vacancy migration volume for chromium and to the bi-vacancy migration for aluminium. For bulk chromium the volume of vacancy migration is $V_{mv} \approx 0.5\,\Omega = 0.6 \times 10^{-29}\,m^3$ (Ω is the atomic volume) for the simple cubic lattice and for aluminium $V_{mv} \approx \Omega = 1.66 \times 10^{-29}\,m^3$ for the face-centred-cubic lattice [5.25]. These results prove the vacancy diffusion nature of the macro-voids. They also allow us to explain the appearance of a minimum in the noise PSD at the maximum void density in the dependence on thickness seen in Figure 5.9.

Alloy films, unlike metal films, have no minimum in the thickness dependence of the noise PSD. Specifically, the intensity of the flicker noise in films of the resistive (metal-silicide) alloys MLT-3M and RS-3001 decreases monotonically with increasing thickness from 20 to 200 nm [5.7]. This phenomenon is due to the absence of the vacancy diffusion mechanism in alloys [5.25], since vacancy movement to sinks, with the formation of macro-voids, is hindered.

Changes in the noise due to annealing or aging

The effect of annealing on the $1/f$ noise has been examined in aluminium films prepared by thermal evaporation in a vacuum [5.28]. The samples were annealed in a vacuum chamber to prevent their oxidation.

The variation of the $1/f$ noise PSD during the course of the thermal treatment was investigated in Al films deposited at 400 K [5.28]. The films were first heated up to 400–450 K over 1200–1800 s, held at this temperature for an equal time, and cooled down to 300–320 K over 3600–5400 s. The PSD of the flicker noise was measured throughout the thermocycle at a fixed frequency and current. Simultaneously, the variation of the film resistance was also measured.

The typical temperature dependence of the $1/f$ noise PSD and the resistance in Al films during a thermocycle are shown in Figure 5.11. Here, S_0 is the $1/f$ noise PSD at $f = 120$ Hz and R_0 is the film resistance prior to annealing ($\Delta R = R - R_0$). The $1/f$ noise increases with heating but drops sharply by 1–3 orders of magnitude after 1200–1800 s at the annealing temperature of 400–410 K. Simultaneously, there is a 2–10% irreversible fall in resistance. During the second (or third) heating cycle (curves 2 and 3 in Figure 5.11), both the noise level and the resistance grow reversibly parallel to the cooling curve in the cycle unless the maximum annealing temperature of the previous cycle is exceeded. Exceeding the maximum annealing temperature of the previous cycle results in a further irreversible decrease in the noise level and the resistance. However, the magnitude of the relaxation-induced decrease in the $1/f$ noise during each cycle is much less than in the previous cycle. The available data indicate that films contain defects with different annealing energies; those with smaller activation energy anneal at a lower temperature.

According to the Matthiessen rule (equation 1.5), the contributions to the resistivity ρ_f in a metal film which originate from the scattering by phonons, defects, and the surface boundaries of the film are additive at

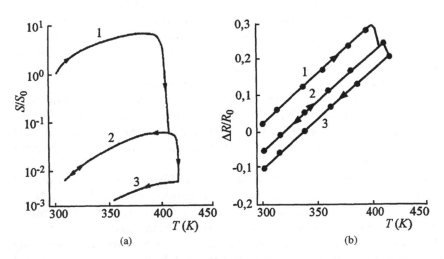

Figure 5.11 The temperature dependence, for Al films during thermocycling [5.28] of: (a) the PSD of the $1/f$ noise and (b) the resistance; $h = 100$ nm, $T_c = 400$ K, $f = 120$ Hz; (1, 2) represents the first cycle and (2, 3) the second cycle.

low concentrations of impurities and defects. Since the contributions of the various types of defects are also additive, it is possible to write [5.23]:

$$\rho_{df} = \frac{mv_F}{Ze^2} \sum_i n_{di} s_i \qquad (5.72)$$

where v_F is an average velocity of a free electron on the Fermi surface; Z is the number of free electrons on each atom; n_{di} is the atomic concentration of the i-th type of defect; s_i is the scattering cross-section of the i-th type of imperfection. The summation in equation 5.72 is carried out over all types of imperfection.

The aluminium films studied in [5.28] had an initial (before annealing) resistivity $\rho_f = 3.8$–$4.2\,\mu\Omega\,cm$ compared with the value for a bulk metal of $\rho_0 = 2.7\,\mu\Omega\,cm$ [5.23]. The linear growth of resistance with temperature (Figure 5.11b) is due to electron scattering by phonons and is the term $\rho_b(T)$ in equation 1.5. The reduced concentration of point defects after annealing has no effect on $d\rho_f/dT$, and only causes a parallel shift in the temperature dependence of the resistance.

Therefore, the irreversible decrease in the resistance and the PSD of flicker noise at 400–410 K can be attributed to the annealing of defects in the crystal lattice. If we assume that, in equation 5.60, $A = 2.2\,\mu\Omega\,cm$ is the contribution of one atomic percent of vacancies to the resistivity of Al [5.23] then we conclude that a 10% fall in the resistivity, caused by annealing, corresponds to a decrease in the vacancy concentration by approximately 0.9 at.%. This appears to be a realistic estimate for metal films (see Section 4.2.1).

The effect on the PSD of the $1/f$ noise of the annealing of induced defects in Al films has been studied [5.29] at low temperature over the range 10 to 300 K. The defects were induced by irradiating the films with electrons of energy 1 MeV with a fluence of 3.7×10^{23} electron m^{-2} at 10 K. The irradiated samples showed a six-fold rise in the noise level but only a 25% increase in resistance (at $T = 10\,K$). Isochronal annealing of the samples for 600 s, with the temperature progressively increased to 300 K, resulted in the recovery of the initial $1/f$ noise level and resistance at 200 K (Figure 5.12). This finding is explained by the lowering of the number of mobile defects. It follows from Figure 5.12 that the various defects have different annealing temperatures.

The effect of γ-radiation on $1/f$ noise in niobium films has been studied [5.30]. Since the activation energy of the induced micro-defects in Nb is higher than in Al, the authors measured the $1/f$ noise at room temperature. The increase in the $1/f$ noise following the irradiation was due to the creation of additional defects in the crystal lattice. Current-induced annealing decreased the noise magnitude to approximately that in the unirradiated samples.

Now, let us consider the changes in the $1/f$ noise intensity with time in naturally ageing films. Figure 5.13 shows the time dependence of the noise at

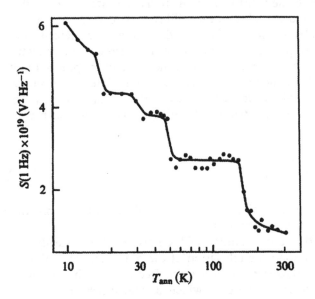

Figure 5.12 The decrease of the PSD ($f = 1$ Hz) of the $1/f$ noise in irradiated Al films at 10 K on isochronal annealing with a gradual rise in temperature [5.29]; the annealing time is 600 s.

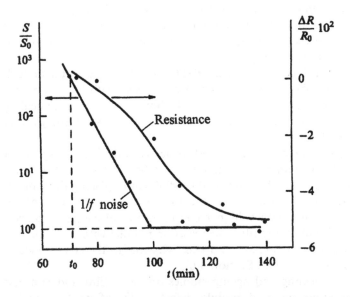

Figure 5.13 The variation of the PSD of the $1/f$ noise at 120 Hz and the resistance with time for Al films [5.7]; $h = 100$ nm, $T_c = 300$ K, R_0 is the film resistance 3500 s after condensation, S_0 is the noise PSD more than 6000 s after condensation.

$f = 120$ Hz together with the resistivity for Al films deposited at $T_c = 300$ K [5.7]. These samples had a highly non-equilibrium structure with a large number of micro-defects. Soon after condensation, the defects were "annealed" at room temperature. This resulted in an irreversible reduction of both the $1/f$ noise and the resistance (time $t = 0$ corresponds to the completion of the film condensation). The $1/f$ noise intensity decreases exponentially with time with a time constant $\tau = 300$ s until $t = 6000$ s. This is characteristic of point micro-defects annealing in accordance with equation 1.19:

$$\Delta S = S - S_0 \propto \exp\left(-\frac{t}{\tau}\right) \tag{5.73}$$

where S_0 is the steady-state value of the noise PSD for times more than 6000 s after the end of the condensation.

Figure 5.13 shows that the relaxation of the noise PSD to a steady-state value requires less time than the relaxation of the resistance. This observation indicates that the flicker noise at a given frequency is due to mobile defects with small activation energies, which anneal faster and at lower temperatures than the defects contributing to the resistance. Defects with higher activation energies contribute to the PSD of $1/f$ noise at lower frequencies.

Similar behaviour of the annealing temperature and the duration of the $1/f$ noise were observed in Cr films deposited at $T_c = 400$ K. In a film deposited on a heated substrate, the relaxation of the $1/f$ noise occurred faster because of the time taken for the substrate to cool from the condensation temperature to room temperature [5.31].

The analysis of the dependence of the PSD of the $1/f$ noise on the annealing time in Cr films showed that it decreases exponentially in accordance with equation 5.73, with a time constant of about 300 s. At an annealing temperature of 620 K almost all the defects with a low activation energy have annealed out over 1800 s.

The frequency exponent, γ, was found to decrease during the ageing of Cr films [5.31]. Measurements taken immediately after fabrication gave $\gamma \approx 2$–3. It decreased to 0.7–1.2 within 6000–9000 s after placing the samples in a vacuum. Higher γ values were found to characterise the condensates further from equilibrium with a higher concentration of excess vacancies. The level of $1/f$ noise in ageing films decreased in parallel with the reduction in the intrinsic mechanical stresses and resistivity due to the coalescence of vacancies into micro-voids (see Section 4.6).

The results of annealing and ageing studies on metal films indicate that they contain different types of mobile defects with different activation energies which hence anneal at different temperatures and with different time constants. These defects contribute to the film resistance in accordance with equation 5.72. Flicker noise in As-deposited metal films contains a component of non-stationary $1/f$ noise induced by excess vacancies, the level of which depends on the concentration of these vacancies. The same component was found in metal films irradiated with high-energy particles, which increases the concentration of micro-defects. Annealing films at temperatures higher than the condensation temperature causes a decrease in the flicker noise due to the decrease in the concentration of vacancies. Annealing of non-equilibrium micro-defects results in a quasi-equilibrium concentration in the films which persists, or changes only very slowly, at temperatures below the annealing temperature. The $1/f$ noise in such films may be regarded as stationary or quasi-stationary.

The temperature dependence of $1/f$ noise

Investigations into the temperature dependence of $1/f$ noise in annealed metal films can help to determine its physical origin. The temperature dependence of $1/f$ noise was first examined by Hooge and Hoppenbrouwers in gold films [5.15], where the dependence of the $1/f$ noise was weaker than $\alpha \propto T^{1/2}$. Voss and Clarke [5.11] reported a decrease in the $1/f$ noise in metal films with decreasing temperature but failed to specify the type of this dependence.

Eberhard and Horn [5.19] appear to have been the first to observe the strong temperature dependence of the $1/f$ noise in Ag, Cu, Au, and Ni films. Later, it was found in Al films [5.28]. In annealed Ag, Au, Cu, and Al films, the temperature dependence of the relative PSD of the $1/f$ noise was shown to be thermally activated over a certain range with an activation energy, E_a:

$$\frac{S_U(f)}{U^2} \propto \exp\left(-\frac{E_a}{kT}\right) \tag{5.74}$$

For Ag, Au, and Cu films, this range lies between 220 and 350 K, while for Al between 220 and 460 K.

The observed values of $E_a = 0.1$–$0.2\,\mathrm{eV}$ for Ag, Cu, and Au films are ascribed to vacancy formation energies although they are significantly lower than those in bulk metals [5.19]. This energy was shown to grow slowly with the film thickness, probably due to the larger grain size. In all the experiments, the activated temperature dependence only held for the $1/f$ noise. The film resistance is dominated by phonon scattering and increased linearly with temperature, in accordance with equation 1.5 and Figure 5.11b.

The temperature dependence of the PSD of the noise has also been determined for Cr, Mo and Ta films [5.7]. The Cr films were prepared by thermal evaporation-condensation in a vacuum on to devitrified glass or oxidized silicon substrates followed by an anneal in a vacuum chamber at $T_{ann} = 620\,\mathrm{K}$ for 600 s. Films of the refractory metals were deposited on oxidized silicon wafers by ion sputtering in the temperature range 300 to 500 K. All the films showed an activation temperature dependence of the

PSD of the noise, which was well approximated by equation 5.74. The experimentally found values of the activation energy corresponded to the energy of one or two bonds in the lattice of a bulk metal. Higher values were recorded for metals with higher atomic binding energies (Mo, Ta), which supports the hypothesis of a vacancy mechanism for the $1/f$ noise in metals.

Figure 5.14 shows the dependence of the PSD of the noise at a frequency of 1 kHz on the inverse temperature for Mo and Ta films. Activation energies derived from these dependencies by the formula:

$$E_a = 2.3k \frac{\Delta \log S}{\Delta(1/T)} \tag{5.75}$$

are: 0.4–0.45 eV for Ta, and 0.3–0.35 eV for Mo.

It has been shown [5.7] that the activation energy is related to the film microstructure. That is, the mean grain size and the degree of crystallinity inside a grain. Higher values are typical of course-grained films. Specifically, an increase of the mean grain size in Cr films from 30–40 to 50 nm led to a rise from 0.2–0.3 to 0.4 eV, with other experimental conditions being identical.

The Chromium films with a small degree of crystallinity inside the grains so that they were practically amorphous showed low activation energy values (0.14 ± 0.02 eV) which increased to 0.33 ± 0.05 eV in films with a moderate

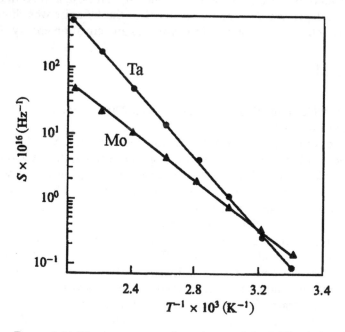

Figure 5.14 The temperature dependence of the PSD of the $1/f$ noise for Ta and Mo films: $h = 0.5\,\mu m$; $f = 10^3$ Hz [5.7].

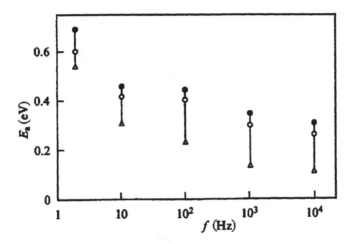

Figure 5.15 The activation energies of the PSD of $1/f$ noise in Mo films of varying thickness at different frequencies [5.7]: (●) 68 nm, (○) 162 nm,
(△) 560 nm.

degree of crystallinity. The mean grain size in the two films was almost
identical at 30–40 nm.

The activation energy has been shown to depend on the frequency at
which the $1/f$ noise PSD is measured [5.7]. The values increased with
decreasing frequency (Figure 5.15). This observation indicates that the $1/f$
noise at different frequencies is due to defects with different activation
energies and the defects with higher activation energies fluctuate slower
and hence contribute to the spectrum at the lower frequencies.

These measurements [5.7] suggest a strong temperature dependence of the
$1/f$ noise due to fluctuations in the number of vacancies in the sample. In
films with a low concentration of mobile defects the $1/f$ noise is generated by
the fluctuations in the mobility caused by the scattering from phonons and
this noise shows a weak temperature dependence [5.15].

The effect of internal macro-stresses on the flicker noise in metal
films

An important argument in support of the vacancy mechanism of flicker noise
in metal films is the dependence of its PSD on mechanical stress, σ found
experimentally in Al, Mo, and Ta films [5.32, 5.33] and obeying the law:

$$S(\sigma) \propto \exp(\sigma V_a/kT) \tag{5.76}$$

where V_a is the activation volume. Figure 5.16 shows the dependence of
the PSD of the noise on the macro-stress for Cr and Mo films, displayed in

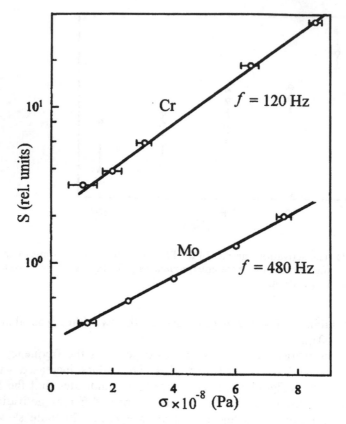

Figure 5.16 The PSD of the $1/f$ noise as a function of the internal mechanical stresses in Cr ($f = 120\,\text{Hz}$) and Mo ($f = 480\,\text{Hz}$) films [5.33].

semi-logarithmic coordinates. The activation volumes can be deduced from the slopes of the straight lines using the equation:

$$V_a = 2.3kT \frac{\Delta \log S(\sigma)}{\Delta \sigma} \qquad (5.77)$$

The dependence of $S(\sigma)$ in equation 5.76 was also found in experiments on Al films obtained by thermal evaporation in a vacuum or an atmosphere of argon [5.32]. However, the activation volumes involved are bigger than the volume per atom due to the peculiarities of the aluminium face centred cubic crystal structure.

It should be emphasised that the dependence $S(\sigma)$ in Figure 5.16 is based on the average value for the internal macro-stresses, σ. Micro-stresses are irregularly distributed across the film and can be significantly, sometimes by an order of magnitude, greater than the macroscopic stress at the grain boundaries. Occasionally, micro-strains can be as large as $\varepsilon \approx 10^{-2}$.

Because micro-stresses are randomly distributed within the bulk of the film and their magnitudes vary locally, the activation energy of vacancy creation and migration is broadened. Assuming Hooke's law, changes in the activation energy due to micro-stresses were calculated using the formula $\Delta E_a = \sigma_m V_a = \varepsilon_m E_{Cr}$ [5.32]. If the micro-strain in a Cr film is $\varepsilon_m = 10^{-2}$, $V_a = 1.2 \times 10^{-29}\,\mathrm{m}^3$, and the Young's modulus $E_{Cr} = 27.3 \times 10^{10}\,\mathrm{Pa}$, then $\Delta E_a \approx 0.2\,\mathrm{eV}$. This value is comparable with the energy per bond in a crystal and accounts for the continuous activation energy spectrum of micro-defects in films from fractions of, to a few electron-volts, which is necessary to obtain the $1/f$ spectrum over a wide frequency range.

Tensile mechanical stresses affect the $1/f$ noise level in metal films through the concentration and mobility of micro-defects. In turn, a change in the concentration of micro-defects during the relaxation processes leads to the alteration of the structural macro-stresses, which increases the elastic strain energy. In this process the energy of the non-equilibrium vacancies decreases because their number decreases. Equilibrium is reached when the thermo-dynamic potential is a minimum. Coalescence of vacancies into voids stops when their motion to the sinks becomes thermodynamically unfavourable. For this reason annealing increases the concentration of vacancies in those films with a high initial value, which accounts for the higher $1/f$ noise in such films.

The internal macro-stresses also affect the spectral shape of the $1/f$ noise. Films with larger stresses have higher γ values [5.32, 5.34]. Figure 5.17 presents the dependence of the frequency exponent γ on the macro-stresses in Cr and Al films [5.32]. The exponent γ grows with an increase in the

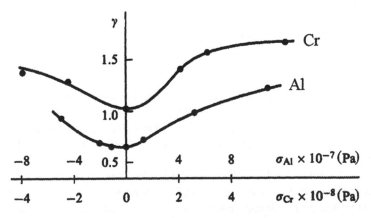

Figure 5.17 The frequency exponent against the internal mechanical stresses in Al and Cr films [5.33].

absolute value of the macro-stresses. These results indicate that different γ values in films of the same material may be due to variations in the level of the internal mechanical stresses.

Dependence of the 1/f noise on mechanical strains

Fleetwood and Giordano [5.35] observed a rise of about one order of magnitude in the $1/f$ noise intensity on the deformation of platinum, gold, silver, lead, and tin films deposited on glass and elastic substrates. Removal of the strain caused the noise to relax to lower levels over a period of a few hours to several months, although the resulting noise magnitude remained higher than initially. The authors explained these changes by the creation and annihilation of structural micro-defects.

Quantitative studies of the effect of externally induced strains and mechanical stresses on the $1/f$ noise in Cr films deposited on glass and Al films deposited on elastic polyimide substrates PM-1 have been reported [5.7, 5.31]. Tensile and compressive stresses were generated in the Cr films by bending the cantilever substrate. Tensile stresses are developed when an external bending force is applied to the free end of the substrate normal to its plane and from the side occupied by the deposited film. The same force applied from the opposite side produces a compressive stress. The displacement of the free end of the substrate was used as a measure of the relative deformation and for calculating the mechanical stress in the elastic strain region [5.31].

Bending the cantilever substrate with an external force generates an asymmetric planar stress in the film. In the case of a tensile force, the internal stresses, σ, and the stress induced by the external force parallel to the x-axis (along the substrate), σ', are added to give the total stress in the film:

$$\sigma_x = \sigma + \sigma'(x) \tag{5.78}$$

In the case of an external compressive force:

$$\sigma_x = \sigma - \sigma'(x) \tag{5.79}$$

When the substrate is fixed as a cantilever, the relative deformation of the film, and hence the mechanical stress induced by the external bending force, is not uniform along the entire length of the film and depends on the distance x from the fixed point. Mean values for the strain and stress in a film at $x = 20\,\text{mm}$ are given below.

Figure 5.18 shows the dependence of the PSD of the noise on the mechanical stresses in the elastic strain regions of two Cr films. Points σ_1 and σ_2 on the x-axis indicate the internal macro-stresses in the absence of any external strains, and the corresponding noise intensities are shown on the y-axis. It can be seen that the film with the larger internal stresses had a higher

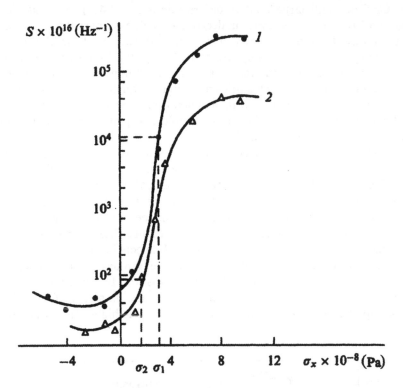

Figure 5.18 The PSD of the $1/f$ noise at 300 Hz against the externally induced mechanical stress in Cr films deposited in an atmosphere of argon ($h = 80$ nm) [5.31]. The argon pressure was; (1) $P_{Ar} = 4 \times 10^{-3}$ Pa and (2) 8×10^{-3} Pa. σ_1, σ_2 are the unstrained internal macro-stresses.

$1/f$ noise level. Increasing the tensile stress to $\sigma \leq 8 \times 10^8$ Pa caused a reversible rise in the magnitude of the $1/f$ noise. The film resistance in this strain region also underwent a reversible increase of 0.5–1%. These dependencies occur in the elastic strain region. An increase in the tensile stress σ_x from 1×10^8 to 3×10^8 Pa lead to an increase in the frequency exponent γ from 1 to 2.5 [5.31].

The application of external compressive forces reduced the noise which reached a low and approximately equal value in all the films which originally had differed in their initial noise level by two or three orders of magnitude, and had been subjected to different mechanical stresses.

Let us discuss the possible causes of the larger $1/f$ noise in the case of elastic strain. The interatomic distance increases as the tensile stresses grow. This leads to a decrease in the activation energy of vacancy formation and a rise in their concentration in accordance with equation 4.5. Essentially all the atoms return to their original positions upon removal of the force that induces the elastic strain. In doing so the vacancy formation energy and $1/f$ noise magnitude take their initial values.

In Cr films with large relative deformations ($\varepsilon \geq 0.4\%$), the noise level (Figure 5.19) and the resistance undergo irreversible changes that show their plastic nature. It should be noted that structural features of the films ensure a large range of elastic strain up to $\varepsilon \approx 0.5$–1% [5.36]. In this case, the strain dependence of the noise and resistance during loading do not coincide with these during unloading. At sufficiently high frequencies ($f > 1\,\text{kHz}$), the PSD of the noise decreases with a relaxation time of 600–3000 s [5.33]. Similar behaviour of the flicker noise of gold and platinum films on deformation has been observed [5.35].

It is known [5.37] that plastic strain is associated with changes in the dislocation structure of a crystalline sample. The movement of dislocations by sliding and climbing are induced by mechanical stresses. On climbing, the dislocations absorb vacancies. This process reduces the number of non-equilibrium vacancies and their aggregates in the crystal and accounts for the lower noise for strains $\varepsilon > 0.5\%$ (see Figure 5.19). This result also confirms that vacancies are the most "noisy" defects in metals.

Partial removal of the external force from the film results in an increase in the vacancy concentration due to their emission by dislocations. This increases the flicker-noise level at low frequencies, as seen in Figure 5.19. In a completely unloaded film, the $1/f$ noise level is higher than the initial.

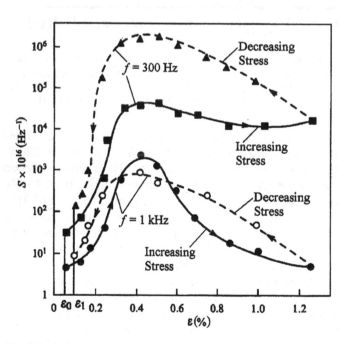

Figure 5.19 The effect of strain on the $1/f$ noise in a Cr film ($h = 80\,\text{nm}$) [5.7]. ε_0 is the initial strain due to internal macro-stresses and ε_1 is the residual strain after unloading.

This is due to the appearance of additional micro-defects. In this case, the film contains a residual strain, ε_1, as seen in Figure 5.19.

In experiments to study the effect of uniform tensile stress on the $1/f$ noise in Al films on an elastic substrate, the stress was increased by applying a force to the free end of the film. The aluminium was deposited at 400 K on a thoroughly cleaned and previously heated substrate. Both the $1/f$ noise and γ increased with increasing tensile stress, as seen in Figure 5.20. The dependence of the increase in the PSD of the noise on the tensile stress obeys the exponential law of equation 5.76.

The effect of structural factors on the level of the 1/f noise

Eberhard and Horn [5.19] reported a 2–6-fold decrease in the $1/f$ noise level after the high temperature anneal of Ag films, which resulted in an approximately 4-fold increase in the mean grain size.

The effect of the microstructure on the $1/f$ noise has been examined in depth for Al, Cr, and Mo films [5.38–5.41]. All these metals showed an increase in the noise with a decrease in the mean grain size.

Figure 5.21 presents the results of a few studies designed to evaluate the effect of structural dispersion on the $1/f$ noise level in Al films [5.38, 5.39]. This is estimated using equation 5.59 for the parameter α, using $n_c = 1.8 \times 10^{23}\,\text{cm}^{-3}$ [5.42]. An increase in the mean grain size reduced the noise magnitude. The parameter α in samples with $d_{av} \approx 200\,\text{nm}$ was close to the commonly observed value $\alpha_H = 2 \times 10^{-3}$ (dashed line). A decrease in the flicker-noise level with increasing grain size has also been observed in Ag

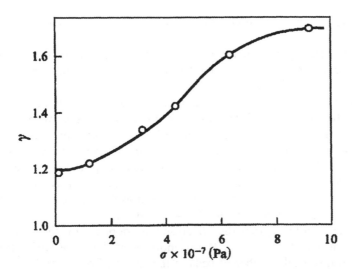

Figure 5.20 The frequency exponent against the tensile stress caused by an external force in an Al film [5.34]; $h = 70\,\text{nm}$.

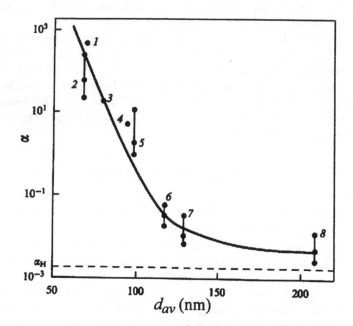

Figure 5.21 The relationship between the $1/f$ noise level and the mean grain size in Al films. Series 1–4 were prepared at different condensation rates [5.38] and series 5–8, at different argon pressures in the chamber [5.39]. Line segments connect α values for different samples of the same set. The dashed line corresponds to the Hooge constant ($\alpha_H = 2 \times 10^{-3}$).

films [5.19], where an α value close to $\alpha_H = 2 \times 10^{-3}$ was achieved for a grain size $d_{av} \approx 200$ nm.

An increase in the $1/f$ noise magnitude with decreasing grain size suggests that the appearance of $1/f$ noise is due to the creation and annihilation of vacancies at grain boundaries.

Let us estimate the vacancy atomic concentration in Al films for different values of the experimental data parameter α in accordance with equation 5.67. Assuming $\rho_f = 3\,\mu\Omega$ cm, $N_0/N_a = 3$, $A = 220\,\mu\Omega$ cm [5.23], and $\ln(f_H/f_L) = 40$, equation 5.67 gives $\alpha \approx 3 \times 10^2\,n_v$. Computed and experimental values for films measured in an atmosphere of argon under different argon pressures in the chamber (see Figure 5.21) agree if $n_v \approx 1 \times 10^{-3} - 3 \times 10^{-5}$ is true of thin films (see Section 4.2.1), with $n_v \approx 10^{-5}$ corresponding to the reference value of $\alpha \approx 2 \times 10^{-3}$.

It has been shown [5.41] that the $1/f$ noise level depends on the vacancy concentration inside the grains, or the proportion of defective lattice cells with respect to the total number of cells in the crystal. For example, in Cr films with approximately similar grain sizes ($d_{av} \approx 30$–40 nm) but differing degrees of crystallinity, the difference in the noise levels amounted to three orders of magnitude. The degree of crystallinity was qualitatively assessed from the spreading and weakening of the interference lines of X-ray and electron diffraction patterns. Cr and Al films with inhomogeneous crystal structures containing a fine-dispersed phase along with large grains exhibited strong flicker noise [5.38].

Films deposited on one substrate by the same technique or even in the same deposition cycle sometimes show different levels of $1/f$ noise. A microstructural examination of such samples revealed impurities at the film surface or various surface defects introduced by the substrate. These defects and admixtures were responsible for a 5–10-fold rise in the noise PSD [5.41] probably because they serve as additional vacancy sources.

Excess noise in metal films with a high concentration of stable defects

Low $1/f$ noise, corresponding to $\alpha_H = 2 \times 10^{-3}$, occurs in high-quality films with a low concentration of mobile defects [5.33, 5.39]. The resistivity of such films is similar to that of bulk metals.

However, metal films with a high concentration of stable defects and a low level of mobile ones also exhibit low $1/f$ noise. Specifically, a noise magnitude of $\alpha \approx 10^{-4}$–10^{-5} was recorded in Cr films obtained by thermal evaporation in an atmosphere of nitrogen [5.30]. These samples had a high concentration of stable defects, and their resistivity, with a temperature coefficient of $\alpha_f \approx 5 \times 10^{-4} \mathrm{K}^{-1}$, was ten times that of bulk chromium.

Figure 5.22 shows the experimental dependence of the PSD of the $1/f$ noise for such a film on the current density squared (curve 1) and the value $\alpha_H = 10^{-3}$ calculated using equation 5.54 with the assumption that $n_0 = 10^{22} \mathrm{cm}^{-3}$ (curve 2). The $1/f$ noise at a current density $j < 5 \times 10^5 \mathrm{A\,cm}^{-2}$ is slightly larger than the thermal noise, whereas for $j > 10^6 \mathrm{A\,cm}^{-2}$ the non-equilibrium resistance fluctuations predominate and the PSD of the $1/f$ noise grows in accordance with the $S \propto j^4$ law. Such a dependence arises from fluctuations of the coefficient R_1 in the expansion of equation 5.46, probably due to local self-heating of the film by the Joule heat resulting from the non-uniform distribution of stable defects (see Section 6.7). The regions of local self-heating give rise to mobile defects, which are responsible for the elevated $1/f$ noise level [5.30].

Stable defects cause a decrease in the relative PSD of the film resistance. Equation 5.67 predicts a decrease in the PSD of the $1/f$ noise induced by fluctuations in the equilibrium vacancy concentration as the resistivity of the film increases due to the additional scattering from stable defects.

Films of Al/Si alloy (Al 1%Si) also have a lower noise magnitude than films of pure Al [5.30] obtained under identical technological regimes. The dimensionless parameters α for Al and the alloy are $\alpha \approx 10^{-2}$ and $\alpha \approx 10^{-3}$, respectively. The low noise level in the Al/Si alloy films compared with that in Al films can be accounted for by a lower concentration of vacancies at the grain boundaries, which are substituted by Si atoms (see Section 7.1.10).

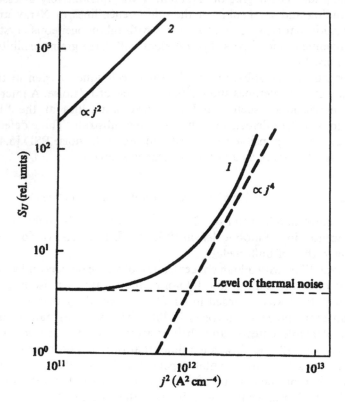

Figure 5.22 The noise PSD of a Cr film deposited in an atmosphere of nitrogen against the current density squared (curve 1) and the results of a computation using the Hooge formula (5.54) (curve 2) [5.30].

Taken together, this result and others discussed earlier in this section indicate that grain boundaries are the principal sources of vacancies in metal films.

5.6   Excess noise in thin-film M1–M2 contacts

This section describes the experimental results for the $1/f$ noise in thin-film Ti–Al contacts. Samples of the contacts were made by thermal evaporation in a vacuum with various thicknesses of the aluminium film on to the Ti in one technological operation by moving a shutter over the substrate during the condensation. The thickness of the Ti film was kept constant at $h_{Ti} = 30$ nm. The condensation temperature was 390 K.

As the noise level of a single contact is small, special samples containing n identical sequentially jointed contacts, –Ti–Al–, were made for the noise measurement. Simultaneously with the manufacture of the thin-film contacts, a homogeneous film of the higher resistance metal (titanium) was deposited on the substrate. It had a length equal to the total length of the metal film in the chain of contacts and an identical width, but contained only two contacts with Al or Cu at the ends. Thus the noise PSD of one contact is determined using the formula:

$$S_c = \frac{S_\Sigma - S_2}{n - 2} \tag{5.80}$$

where S_Σ is the noise PSD of a chain consisting of n contacts; S_2 is the noise PSD of the two-contact element.

As well as the noise, the contact resistance was measured to evaluate the thin film resistance and density of macro-voids. Samples of the contact structures were measured immediately after manufacture and again after annealing in a vacuum for 900 s at a temperature of 520 K.

Figure 5.23 shows the dependence of the PSD of the $1/f$ noise on the Al film thickness for a Ti–Al contact with a Ti film thickness of 30 nm. It is clear that the annealing resulted in a decrease in the noise level and a change in the thickness dependence of the PSD for the Al films. The minimum, corresponding to the lowest vacancy concentration, was associated with an Al-film thickness of $h_{Al} \approx 80$ nm and occurred when the resistance was a minimum and the macro-void density a maximum. These results are similar to those presented in Figure 5.9 for continuous Al films.

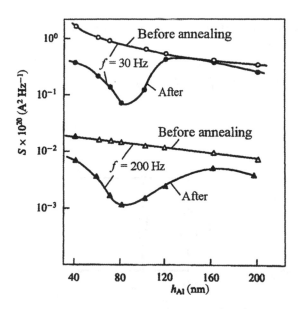

Figure 5.23 The $1/f$ noise PSD of a Ti–Al thin-film contact against the Al film thickness ($h_{Ti} = 30$ nm) before and after annealing [5.7].

For the flicker noise in the contacts before annealing, $\gamma \approx 2$–3, whereas for Ti films it was $\gamma \approx 1.0$–1.2. These higher values for contacts that had not been annealed suggest the appearance of non-equilibrium $1/f^\gamma$ noise in these systems.

5.7   Noise equivalent circuits

For the analysis of noise in electric circuits it is convenient to replace the actual noisy element by a noiseless one with the same resistance together with a noise voltage generator connected in series with it. Alternatively, the noise equivalent circuit could be a noise current generator in parallel with a conductance that is equal to the conductance of the noise source.

Equivalent circuits for the noisy resistor are shown in Figure 5.24. The noise voltage generator $U_n(t) = U_T(t) + U_{exc}(t)$ is represented by the sum of the thermal and excess noises, connected in series with the resistance. The current generator $I_n(t) = I_T(t) + I_{exc}(t)$ represents the sum of the thermal and excess noise, connected in parallel with a conductance $G = 1/R$. The thermal noise PSD for a voltage source is given by equation 5.18a and for a current source by equation 5.18b.

The PSD of the excess noise is given by equation 5.39. The mean square of the voltage fluctuation of the equivalent noise generator (Figure 5.24a) in the frequency band from f_1 to f_2 is equal to:

$$\overline{U_n^2} = \overline{U_T^2} + \overline{U_{exc}^2} = 4kTR\Delta f + \int_{f_1}^{f_2} K_1 I^a f^{-\gamma} \, df \tag{5.81}$$

and the mean square of the current fluctuation of the equivalent noise generator (Figure 5.24b) is

$$\overline{I_n^2} = \overline{I_T^2} + \overline{I_{exc}^2} = 4kTG\Delta f + G^2 \int_{f_1}^{f_2} K_1 I^a f^{-\gamma} \, df \tag{5.82}$$

where $\Delta f = f_2 - f_1$.

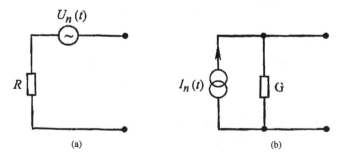

(a)   (b)

Figure 5.24 Equivalent circuits of a noisy element (a) with a voltage generator and (b) with a current generator.

Equivalent noise resistance

Sometimes the noise generated in an active element of an electrical circuit, or in an amplifier, is represented by an equivalent resistor generating the same amount of thermal noise [5.1]. It is assumed that the temperature of it is fixed at room temperature, T_0. Then the equivalent noise resistance is determined from the Nyquist formula 5.18 and is equal to:

$$R_N = \frac{\overline{U_{NA}^2}}{4kT_0\Delta f} \tag{5.83}$$

where $\overline{U_{NA}^2}$ is the total mean square of the noise voltage of the amplifier from all sources in a bandwidth Δf, referred to the input.

Thus it is possible to assume that the amplifier does not make a noise, and the noise measured at its output is created by an equivalent noisy resistor inserted at the input of the amplifier. The magnitude of the equivalent noise resistance will depend on both the measurement frequency and on the mode of operation of the element. At low frequencies the noise of the amplifier has a $1/f$ spectrum. At frequencies higher than several or tens of kilo-Hertz the noise spectrum normally becomes constant. Note that the equivalent noise resistor does not exist as an element of the circuit. It is not possible to speak, for example, about the magnitude of any voltage drop across it or about the power dissipated in it. The noise resistance is introduced only to characterise the noise properties of the circuit elements and amplifiers.

5.8 The measurement of noise in conducting films

The power spectral density of a noise signal is measured with a spectrum analyser (SA) which measures the spectral intensity of a signal at different frequencies. A basic spectrum analyser may be considered to consist of a frequency selective narrow-band filter, which separates a noise spectrum into its frequency components. In Figure 5.25 a block diagram of a basic spectrum analyser is given.

The pre-amplifier generates as little noise as possible. For the narrow-band filter piezoelectric or resonant LC-filters can be used. After an output amplifier the noise signal is passed to a detector with an RC-filter, which produces rectification and time-averaging of the noise voltage signal. An indicator serves as a measure of the noise power. The mean square voltage of the noise can be measured by a square-law voltmeter. For the spectral analysis of a noise in the required frequency band it is necessary to switch the frequency of the narrow-band filter, or use a parallel set of narrow-band filters adjusted at different frequencies. A spectrum analyser operating as a superheterodyne receiver covers the widest frequency range.

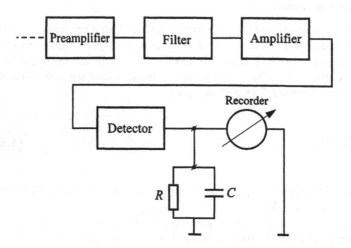

Figure 5.25 The block diagram of a basic spectrum analyser.

If the narrow-band amplifier is adjusted to a centre frequency f_0, with a band width Δf, and measures a noise with a power spectral density $S(f)$, then the mean square voltage of the noise at the output of the amplifier is:

$$\overline{U^2} = \int_{f_0-\Delta f/2}^{f_0+\Delta f/2} K^2(f)S(f)\mathrm{d}f \tag{5.84}$$

where $K(f)$ is the voltage gain of the amplifier which gives $\overline{U^2} = K^2(f_0)S(f_0)\Delta f$.

Thus we can express the noise PSD at the frequency f_0 in terms of $\mathrm{V^2\,Hz^{-1}}$ through the mean square noise voltage:

$$S(f_0) = \frac{\overline{U^2}}{K^2(f_0)\Delta f} \tag{5.85}$$

For a more exact definition of $S(f_0)$ the band width Δf should be as narrow as possible. The bandwidth is the smallest frequency interval in which two close components of a noise spectrum can be resolved. However, the transmission band cannot be made too narrow since the relative error in a measurement of the noise intensity is governed by the fluctuating nature of the signal. If an RC-filter is used to average the signal after the detector the relative error of a noise measurement is determined by the formula [5.3]:

$$\delta\% = \frac{100\%}{\sqrt{2\Delta f\tau}} \tag{5.86}$$

where $\tau = RC$ is the time constant of the RC-filter.

The error given in equation 5.86 is because of the inevitable fluctuation of a noise signal about the average value. To decrease the measurement error it is necessary to increase the time constant of the averaging filter, but then the measurement time is also increased. Thus, an increase in the accuracy of the measured noise by a decrease in the transmission band can only be achieved at the expense of an increase in the measurement time.

With the development of computers, at all but the highest frequencies, spectrum analysers with analogue filters have been supplanted by analysers with digital filters for the signals. The amplified signal is digitised and then the processing is carried out by software. Digital signal analysers have several advantages compared with analogue ones. Usually the spectral density of a random process is determined by calculating the auto-correlation function through the Fourier transform equation 5.6. The spectral calculation is carried out on the signal captured during a finite period of time. The digitisation rate determines the maximum frequency in the spectrum and the length of the sample determines the lowest frequency. To reduce the error in the calculation of the PSD of the random process due to the finite length of the sample of the signal the calculation is made with some weighting function over the sample length called the spectral window. These windows are given in the literature and their use allows the calculation of the noise PSD with high accuracy. Normally many separate measurements of the PSD are averaged.

A block diagram of an experimental apparatus for the measurement of the noise PSD of film structures is shown in Figure 5.26. The $1/f$ noise voltage arises from the resistance fluctuations of the sample, R_x, if a direct current is passed through it from the current source.

In order to eliminate the excess noise of the contacts, a four-probe method of measurement is used with a specially constructed thin-film resistor with

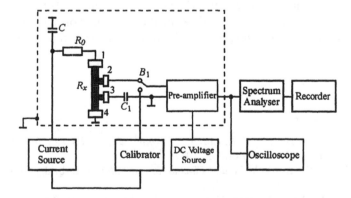

Figure 5.26 The block diagram of an apparatus to measure the noise in conducting films.

four contacts. A direct current source is connected to the end contacts (1,4), and the noise voltage is measured between two central contacts (2,3). The current does not pass through these so there is no current noise due to any fluctuations in the contact resistances.

The current supply and the pre-amplifier are powered from alkaline batteries or from some other power supply with a low level of noise. If the current source is replaced by a voltage source then for a constant current through the sample $R_0 \gg R_x$. Resistor R_0 is wire-wound so that it has low excess noise. The noise voltage from the conducting film is amplified by the low-noise pre-amplifier and then passed to the spectrum analyser. An oscilloscope is connected in parallel with the input of the spectrum analyser to monitor the size and shape of the signal being analysed.

To eliminate pickup from the rest of the electrical system or from other external electric fields the sample and pre-amplifier are carefully shielded. The system is calibrated with a noise generator, a sinusoidal oscillator or by using the thermal noise of a set of resistors with no current flowing.

To measure the excess noise it is necessary to take into account the thermal noise of the sample and the noise of the pre-amplifier. The excess noise PSD is calculated by the formula:

$$S_{\text{exc}}(t) = \frac{\overline{U^2(f)} - \overline{U_T^2(f)}}{K_0^2 K^2(f)\Delta f} \tag{5.87}$$

where $\overline{U^2(f)}$ and $\overline{U_T^2(f)}$ are the mean squares of a noise voltage in a narrow frequency band near the measurement frequency measured when a direct current I_0 is passed through the sample and without current, respectively; $(U_T^2(f)$ is the effective voltage of the thermal noise of the film and the noise of the pre-amplifier); $K(f)$ is the gain of the measuring system at the frequency f; Δf is the effective band width of the spectrum analyser; $K_0 \approx R_x/(R_0 + R_x)$ is the gain of the sample bias circuit.

For noise measurements at low frequencies it is recommended that frequencies are selected that are not multiples of the frequency of the electrical power supply network (50 or 60 Hz) in order to eliminate pickup from that source.

Literature for Chapter 5

5.1 Lukianchikova, N.B., *Noise Research in Semiconductor Physics* (ed. B.K. Jones), Gordon and Breach, London, 1997.

5.2 Kogan, Sh.M., *Electronic Noise and Fluctuations in Solids*, Cambridge University Press, 1996.

5.3 van der Ziel, A., *Noise, Sources, Characterisation, Measurement*, Prentice-Hall, Englewood Cliffs, 1970.

5.4 Hooge, F.N., Kleinpenning, T.G.M. and Vandamme, L.K.J., "Experimental studies on 1/f noise", *Rep. Prog. in Physics*, **44** (1981) 479–532.

5.5 Dutta, P. and Horn, P.M., "Low-frequency fluctuations in solids: 1/f noise", *Rev. Mod. Phys.*, **53** (1981) 497–516.

5.6 Weissman, M.B., "1/f noise and other slow, non-exponential kinetics in condensed matter", *Rev. Mod. Phys.*, **60** (1988) 537–571.

5.7 Zhigal'skii, G.P., "1/f noise and non-linear effects in metal films", *Soviet Physics-Uspekhi*, **40** (1997) 599–622.

5.8 Jones, B.K., "Electrical noise as a measure of quality and reliability in electronic devices", *Advances in Electronics and Electron Physics*, **87** (1994) 201–257.

5.9 Uren, M.J., Day, D.J. and Kirton, M.J., "1/f noise and random telegraph noise in silicon metal-oxide-semiconductor field-effect transistors", *Appl. Phys. Lett.*, **47** (1985) 1195–1197.

5.10 Zhigal'skii, G.P., "Non-equilibrium flicker noise in conducting films", *Russian J. Phys. Chem.*, **69** (1995) 1218–1220.

5.11 Voss, R.F. and Clarke, J., "1/f noise from systems in thermal equilibrium", *Phys. Rev. Lett.*, **36** (1976) 42–45.

5.12 Francis, J.D. and Jones, B.K., "Direct correlation between 1/f and other noise sources", *J. Phys. D.*, **8** (1975) 1172–1174.

5.13 Keshner, M.S., "1/f noise", *Proc. IEEE*, **70** (1982) 212–218.

5.14 Hooge, F.N., "1/f noise sources", *IEEE Trans. Electron. Dev.*, **41** (1994) 1926–1935.

5.15 Hooge, F.N., "The relation between 1/f noise and the number of electrons", *Physica B*, **162** (1990) 344–352.

5.16 Hooge, F.N., "40 years of 1/f noise modelling", *Proc. 14th Int. Conf. on Noise in Physical Systems and 1/f Fluctuations*, Leuven, 1997, World Scientific, pp. 3–10.

5.17 Kleinpenning, T.G.M., "1/f noise in p-n diodes", *Physica*, **B98** (1980) 289–299.

5.18 Musha, T., Borbely, G. and Shoji, M., "1/f phonon-number fluctuations in quartz observed by laser light scattering", *Phys. Rev. Lett.*, **64** (1990) 2394–2397.

5.19 Eberhard, J.W. and Horn, P.M., "Temperature dependence of 1/f noise in silver and copper films", *Phys. Rev. Lett.*, **39** (1977) 643–646.

5.20 Robinson, F.N.H., "A mechanism for 1/f noise in metal", *Phys. Lett.*, **A97** (1983) 162–163.

5.21 Pelz, J. and Clarke, J., "Dependence of 1/f noise on defects induced in copper films by electron irradiation", *Phys. Rev. Lett.*, **55** (1985) 738–741.

5.22 Pelz, J. and Clarke, J., "Quantitative 'local-interference' model of 1/f noise in metal films", *Phys. Rev.*, **B36** (1987) 4479–4482.

5.23 Thompson, M., *Defects and Radiation Damage in Metals*, Cambridge University Press, Cambridge, 1969.

5.24 Celasco, M., Fiorillo, F. and Mazzetti, P., "Thermal-equilibrium properties of vacancies in metals through current-noise measurements", *Phys. Rev. Lett.*, **36** (1976) 38–40.

5.25 Bokshtein, B.S., *Diffuziya v Metallakh (Diffusion in Metals)* (in Russian), Metallurgiya, Moscow, 1978.

5.26 Fleetwood, D.M., Mosden, J.T. and Giordano, N., "1/f noise in platinum films and ultrathin platinum wires: evidence for a common, bulk oxygen", *Phys. Rev. Lett.*, **50** (1983) 450–453.

5.27   Celasco, M., Fiorillo, F. and Masoero, A., "Comment on $1/f$ noise and its temperature dependence in silver and copper", *Phys. Rev.*, **B19** (1979) 1304–1306.

5.28   Zhigal'skii, G.P. and Bakshi, I.S., "Excess noise in thin aluminium films", *Radio Eng. Electron. Phys.*, **25** (1980) 61–68.

5.29   Briggmann, J., Dagge, K., Frank, W., Seeger, A., Stoll, H. and Verbruggen, A.H., "$1/f$ noise in low-temperature-irradiated aluminium films", *Proc. AIP Conf. 285 on Noise in Physical Systems and $1/f$ Fluctuations*, St. Louis, 1993, 607–610.

5.30   Potemkin, V.V., Stepanov, A.V. and Zhigal'skii, G.P., "$1/f$ noise in thin metal films: the role of steady and mobile defects", *Proc. AIP Conf. 285 on Noise in Physical Systems and $1/f$ Fluctuations*, St. Louis, 1993, 61–64.

5.31   Zhigal'skii, G.P., Kurov, G.A. and Siranashvili, I.Sh., "Excess noise and mechanical stress in thin chromium films", *Radiophys. Quantum Electron.*, **26** (1983) 162–166.

5.32   Zhigal'skii, G.P., "An effect of structure factors and mechanical stress on $1/f$ noise in metal films", *Proc. AIP Conf. 285 on Noise in Physical Systems and $1/f$ Fluctuations*, 1993, St. Louis, 81–84.

5.33   Zhigal'skii, G.P., "Relationship between $1/f$ noise and non-linearity effects in metal films", *JETP Lett.*, **54** (1991) 513–516.

5.34   Zhigal'skii, G.P., Sokov, Yu.E. and Tomson, N.G., "Investigation of the $1/f$ noise dependency in thin aluminium films on internal mechanical stresses", *Radio Eng. Electron. Phys.*, **24** (1979) 137 [*Radiotekh. Elektron.*, **24** (1979) 410–413].

5.35   Fleetwood, D.M. and Giordano, N., "Effect of strain on the $1/f$ noise of metal films", *Phys. Rev.*, **B28** (1983) 3625–3627.

5.36   Hoffman, R.U., "Mechanical property of thin condensed films", *Physics of Thin Films*, Hass, G. and Thun, R. (eds), Academic Press, New York and London, **3** (1966) 211–273.

5.37   Hirth, J. and Lothe, J., *Theory of Dislocations*, McGraw-Hill, New York, 1968.

5.38   Andrushko, A.F., Bakshi, I.S. and Zhigal'skii, G.P., "Effects of structural factors on the $1/f$ noise of aluminium films", *Radiophys. Quantum Electron.*, **24** (1981) 343–346.

5.39   Potemkin, V.V., Bakshi, I.S. and Zhigal'skii, G.P., "$1/f$-type noise in polycrystalline aluminium films", *Radio Eng. Electron. Phys.*, **28** (1983) 89 [*Radiotekh. Elektron.*, **28** (1983) 2211–2216].

5.40   Zhigal'skii, G.P., Markaryants, E.A. and Fedorov, A.S., "Investigation of electro-physical properties of the refractory metal films manufactured by ionic sputtering", *Poverkhnost': Fizika, Khimiya, Mekhanika.*, 1993, 78–86.

5.41   Zhigal'skii, G.P., Karev, A.V., Siranashvili, I.Sh., Andrushko, A.F. and Kovalev, V.D., "An effect of structure factors on $1/f$ noise of small grain chromium films", *Radio Eng. Electron. Phys.*, **33** (1990) 870–873 [*Radiotekh. Elektron.*, **33** (1990) 1181–1184].

5.42   Kittel, C., *Introduction to Solid State Physics*, J. Wiley, New York, 1967.

6 Non-linear conduction and non-equilibrium noise in thin films

6.1 Introduction

A non-linearity of the current–voltage characteristic (CVC), or deviation from Ohm's law, is observed in nearly all of the elements of integrated circuits. The resistance of a conducting film is then a non-linear function of the current and can be represented as [6.1]:

$$R(I,t) = \sum_{n=0}^{\infty} R_n(t) I^n(t) \tag{6.1}$$

Here, the time dependence of the coefficients in the $R_n(t)$ series represents the fluctuations in the film resistance. The fluctuations of the $R_0(t)$ term represent the equilibrium $1/f^\gamma$ noise. The thermodynamic equilibrium $1/f$ noise in metal films has been described in Chapter 5.

The other terms of the series in equation 6.1 ($n \geq 1$) give the non-linearity and correspond to various trapping and scattering processes of the charge carriers. These can be the scattering by mobile defects and the passage above and through the barriers in granular films (see Chapter 2). The latter can also take place in films when thin dielectric layers of oxides or nitrides are formed at the boundaries of the grains (see Section 2.2).

These higher coefficients in the series do not contribute to the resistance fluctuations when $I = 0$. Since the fluctuations of these coefficients appear only when $I > 0$ they can be related to non-equilibrium fluctuations in the conductivity of the film.

The size of the CVC non-linearity and the non-equilibrium fluctuations are valuable quantities for evaluating the quality of materials and the reliability of metal films and resistors [6.2, 6.3].

In this chapter the theoretical analysis of the CVC non-linearity and non-equilibrium noise in conducting films is considered and some experimental results are described.

6.2   The spectrum of the response to sinusoidal excitation

When a sinusoidal current with amplitude I_1 and frequency f_1 (angular frequency $\omega_1 = 2\pi f_1$):

$$I(t) = I_1 \sin \omega_1 t \tag{6.2}$$

flows through a film it produces a response in the voltage because of the non-linearity of the CVC so that:

$$U(t) = R(I, t) I_1 \sin \omega_1 t \tag{6.3}$$

Using equation 6.1 the voltage drop across the sample can be written as:

$$U(t) = U_0(t) + \sum_{n=1}^{\infty} U_n(t) \sin n\omega_1 t \tag{6.4}$$

where $U_0(t)$ is the voltage of the direct component (zero harmonic) and $U_n(t)$ ($n = 1, 2, 3, \ldots$) are the amplitudes of the nth voltage harmonics. $U_0(t)$ and $U_n(t)$ are random functions of time corresponding to the fluctuations of the $R_n(t)$ coefficients in equation 6.1.

For a weakly non-linear element like a metal film the terms decrease rapidly with n so that for ordinary working currents:

$$\overline{R_0} \gg \overline{R_n} I_1^n \gg \overline{R_{n+1}} I_1^{n+1} \tag{6.5}$$

Here $\overline{R_0}$, $\overline{R_n}$ and $\overline{R_{n+1}}$ are the time-averaged coefficients $R_0(t)$, $R_n(t)$ and $R_{n+1}(t)$:

$$\overline{R_n} = \lim_{T \to \infty} \frac{1}{T} \int_0^T R_n(t) \mathrm{d}t \quad (n = 0, 1, 2, \ldots) \tag{6.6}$$

For most practical applications, it is sufficient to consider the first few terms in the expansion. Then the voltage of the fundamental frequency signal at f_1 (first harmonic) across the film resistor is:

$$U_1(t) = R_0(t) I_1 + \frac{3}{4} R_2(t) I_1^3 \approx R_0(t) I_1 \tag{6.7}$$

The amplitudes of the direct voltage component, U_0, the second, U_2, and third, U_3, harmonics in equation 6.4 are expressed as a series of coefficients as:

$$U_0(t) = \frac{1}{2} R_1(t) I_1^2 \tag{6.8}$$

$$U_2(t) = \frac{1}{2} R_1(t) I_1^2 \tag{6.9}$$

$$U_3(t) = \frac{1}{4} R_2(t) I_1^3 \tag{6.10}$$

The so-called $1/\Delta f$ noise [6.4], which represents the random amplitude modulation of the fundamental frequency signal $U_1(t)\sin\omega_1 t$, is generated by fluctuations of the linear resistance $R_0(t)$. Fluctuations of the $R_1(t)$ and $R_2(t)$ coefficients in equation 6.1 (the quadratic and cubic CVC terms) lead to the modulation by the $1/f$ noise of the second and third harmonics amplitudes, $U_2(t)$ and $U_3(t)$, of the response signal in accordance with equations 6.9 and 6.10.

When considering the CVC, the quantities $R_0(t)$, $R_1(t)$ and $R_2(t)$ in equations 6.7–6.10 can be changed to the time-average values of the coefficients in the series 6.1 by using equation 6.6.

It can be seen from equations 6.8 and 6.9 that the quadratic non-linearity of CVC causes the appearance of the zero-order and second harmonics. The amplitudes are proportional to I_1^2. The cubic non-linearity of the CVC causes the third harmonic in the voltage output, which is proportional to I_1^3. In metallic films the cubic non-linearity can result from the Joule heating of the film caused by the scattering of carriers by the lattice and mobile defects [6.1].

6.3   Experiments to study non-linear effects in films

To determine the deviation of the CVC from Ohm's law for metal films and other conducting layers by the direct measurement of the resistance using a direct current is impossible because of the small level of the CVC non-linearity. Measurement of the CVC non-linearity may be made using an alternating current by measuring the amplitudes of the harmonics. This method allows the measurement of a very small CVC non-linearity.

The method is shown in Figure 6.1. A pure sinusoidal current signal at the fundamental frequency is passed through the sample, R_x. It passes from a signal generator through a low-pass filter to suppress any higher harmonics. The sinusoidal current, $I_1\sin\omega_1 t$, flowing through the sample generates a voltage at the third harmonic, $U_3\sin 3\,\omega_1 t$, because of the cubic non-linearity. This signal-response is passed by a high-pass filter to remove the exciting signal and any second harmonic, and then detected by a selective microvoltmeter. This latter is also used for the measurement of the amplitude, U_1, of the test-signal, $U_1\sin\omega_1 t$, across the sample.

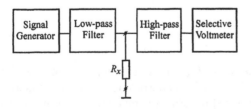

Figure 6.1 The block diagram of the circuit for the measurement of the level of CVC non-linearity of a film structure: R_x is the sample.

The bias of the sample is produced by a direct current, $I_0 = U_0/R_x$. Then the full current flowing through the specimen is:

$$I(t) = I_0 + I_1 \sin \omega_1 t \tag{6.11}$$

The amount of non-linearity within the measuring device is determined by replacing the sample by a wire-wound resistor with a resistance equal to the resistance of the sample. The amplitude of the third harmonic of the signal-response is calculated using the equation:

$$U_3 = \frac{U_{3Rx} - U_{3R0}}{K_0} \tag{6.12}$$

where U_{3Rx} is the amplitude of the third harmonic measured by the selective micro-voltmeter which includes the total non-linearity of the film sample and measuring device; U_{3R0} is the amplitude of the third harmonic when the wire resistor is connected instead of the sample; K_0 is the gain experienced by a voltage from a source of resistance R_x applied to the input of the selective micro-voltmeter.

The degree of CVC non-linearity is characterised by the non-linearity coefficients defined as the ratio of the third harmonic amplitude, U_3, to the amplitude of the fundamental signal, $U_1 = R_x I_1$:

$$K_{n3} = U_3/U_1 \tag{6.13}$$

or expressed in decibels [6.2]:

$$K_{NL} = 20 \log(U_3/U_1) \tag{6.14}$$

Sometimes, the non-linearity coefficient is presented as:

$$K_{N3} = U_3/I_1^3 \tag{6.15}$$

or in the form:

$$K_{N3} = U_3/U_1^3 = U_3/R_0^3 I_1^3 \tag{6.16}$$

In this form, from equation 6.10, K_{N3} does not depend on the signal amplitude I_1, so that samples, for which the harmonics U_3 were measured at different currents, can be compared in terms of the degree of non-linearity.

By analogy with equations 6.12–6.16, similar relations can be written for the CVC non-linearity coefficient of the second harmonic.

6.4 The effect of different variables on the CVC non-linearity in metal films

There is found to be a very similar dependence of the third harmonic amplitude, U_3, and the flicker noise intensity in metal films with various technological factors such as the thickness, argon pressure in the vacuum chamber during condensation and the bias voltage across the substrate during film preparation by the ion sputtering technique. The minimum value of U_3, with the thickness of a Cr film [6.5], is close to the minimum of the PSD of the $1/f$ noise shown in Figure 5.9 at a film thickness of 80–100 nm.

Figure 6.2 shows the dependence on thickness of the non-linearity coefficients (equation 6.13) for chromium films deposited in a vacuum (curve 1) and an atmosphere of argon ($P_{Ar} = 7 \times 10^{-3}$ Pa) (curve 2). Measurements were made at the same alternating current through samples of different thicknesses. The lowest K_{n3} value is observed in the thickness range between 60 and 120 nm. The highest macro-void density and minimum resistivity and $1/f$ noise PSD also occur within this range. These observed effects can be explained since films of thickness $h \approx 100$ nm have the lowest concentration of vacancies (see Section 4.3).

Alloy films, unlike metal films, have no minimum in the thickness dependence of the CVC non-linearity coefficients or the noise PSD. This is because there is little vacancy diffusion in alloys, so that the vacancy movement to sinks with the formation of macro-voids is small.

Chromium films fabricated in an atmosphere of argon showed a minimum in both the PSD of the $1/f$ noise [6.6] and the non-linearity coefficient for an argon pressure of 10^{-2} Pa in the deposition chamber (Figure 6.3).

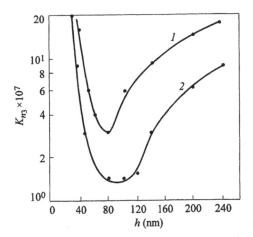

Figure 6.2 The current–voltage characteristic non-linearity coefficient (equation 6.13) against thickness for chromium films deposited in; (1) vacuum and (2) an atmosphere of argon ($P_{Ar} = 7 \times 10^{-3}$ Pa) [6.5]: $f_1 = 10$ kHz.

The temperature dependence of the CVC non-linearity coefficient have been investigated in Mo and Ta films subjected to a weak sinusoidal signal with frequency $f_1 = 10\,\text{kHz}$ and a power $P \leq 10\,\mu\text{W}$ [6.7] (Figure 6.4).

Films with a large $1/f$ noise level showed a thermally activated dependence on temperature of the third harmonic amplitude, U_3 [6.7, 6.8]. The activation energies derived from the temperature dependence of the non-linearity coefficient for Mo and Ta films using the equation:

$$E_a = 2.3\,k\Delta \log U_3/\Delta(1/T) \tag{6.17}$$

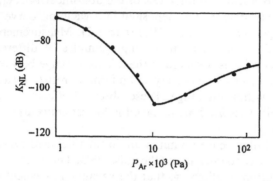

Figure 6.3 The current–voltage characteristic non-linearity coefficient (equation 6.14) for chromium films as a function of the argon pressure in the deposition chamber [6.6]: $h = 0.080\,\mu\text{m}$, $f_1 = 10\,\text{kHz}$, $U_1 = 200\,\text{mV}$.

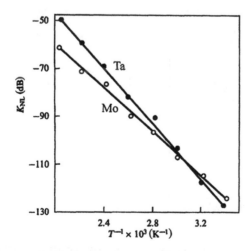

Figure 6.4 The current–voltage characteristic non-linearity coefficient for Ta and Mo films against inverse temperature [6.7]: $h = 0.5\,\mu\text{m}$, $f_1 = 10\,\text{kHz}$, $U_1 = 250\,\text{mV}$.

are $E_a = 0.3$–0.35 eV and $E_a = 0.4$–0.45 eV respectively and coincided with the activation energies for the PSD of the $1/f$ noise at frequencies $f > 10^3$–10^4 Hz, measured from the temperature dependence of $1/f$ noise (see Figure 5.14).

6.5   Mechanisms of cubic non-linearity in metal films

It was shown earlier that the third harmonic amplitude, the non-linearity coefficient, and the flicker noise intensity in metal films depend in similar ways on various technological factors. Chromium films fabricated in an atmosphere of argon showed a minimum in both the $1/f$ noise and non-linearity coefficient when deposited at an argon pressure of 10^{-2} Pa. Also the temperature dependencies of the $1/f$ noise intensity and of the current–voltage characteristic non-linearity coefficient were shown to be thermally activated with equal values of the activation energy at high enough frequencies ($f \geq 1$ kHz).

These findings suggest a similar physical origin for the mechanisms responsible for the $1/f$ noise and the cubic non-linearity of the current–voltage characteristic for metal films. It is possible that the non-equilibrium vacancies are one source of the non-linearity in metal films.

It was shown in Section 5.5 that the $1/f$ noise in metal films with an elevated concentration of mobile defects is caused by the creation and annihilation of vacancies. Thus the spectral density of the $1/f$ noise is proportional to the quasi-equilibrium vacancy concentration (equation 4.5) in the bulk of the films and is defined by equation 5.66.

Now we discuss the physical processes that bring about non-linear effects in continuous metal films. We will consider the mechanism for the origin of the third harmonic voltage, which arises when a sinusoidal current flows through a film [6.8].

The temperature of the film above the equilibrium temperature can be defined through the thermal resistance to the substrate, R_T, and the power, P, dissipated by the signal:

$$\Delta T = T - T_0 = R_T P = (R_T L \rho_f / bh) I_1^2 \sin^2(\omega_1 t) \tag{6.18}$$

Here L, b and h are the length, width and thickness of the film, respectively and ρ_f is the film resistivity. It is assumed here that the thermal relaxation time of the film-substrate system is short compared with the period of the heating, which is true for films deposited on oxidised silicon and measured at a signal frequency $f_1 < 30$ kHz. The effects of thermal inertia on the third harmonic amplitude lead only to a change in the phase and absolute amplitude of the third harmonic, which is not important for the study of the non-linearity mechanism in thin metal films.

A change in the film temperature induced by a sinusoidal signal given by equation 6.18 leads to periodic changes in the vacancy concentration in the film according to equation 4.5 and hence the film resistivity and the resistance of the sample. Assuming that only vacancies with an activation energy E_V contribute to the part of the specific resistance related to mobile defects and using equations 1.5, 4.5 and 5.60, the temperature dependence of the film resistivity can be written as:

$$\rho_f(T) = \rho_{01}(1 + \alpha_f \Delta T) + AA_v \exp[-E_v/k(T_0 + \Delta T)] \tag{6.19}$$

Here, ρ_{01} is the total film resistivity at T_0 due to scattering from phonons ρ_b, fixed defects ρ_1 and the film boundaries ρ_s:

$$\rho_{01} = \rho_b + \rho_1 + \rho_s \tag{6.20}$$

α_f is the temperature coefficient of resistivity (TCR) of the film ($\alpha_f > 0$) defined as above and equal to:

$$\alpha_f = \frac{\alpha_0 \rho_b(T_0)}{\rho_{01}} \tag{6.21}$$

where α_0 is the temperature coefficient of resistance of the bulk metal.

The first and second terms in equation 6.19 define the contribution of the scattering from phonons and vacancies to the film resistance, respectively and E_v is the lowest activation energy of mobile defects.

The voltage drop across the film is:

$$U = \rho_f K I_1 \sin \omega_1 t \tag{6.22}$$

where $K = L/bh$ is the dimensional coefficient.

Substitution of equations 6.18 and 6.19 into equation 6.22 and expanding in powers of ΔT while assuming that $E_v \Delta T/kT_0^2 \ll 1$ ($E_v \Delta T/kT_0^2 = 0.06$ for $E_v = 0.5\,\text{eV}$ and $\Delta T = 1\,\text{K}$) gives an equation for the third harmonic:

$$U_3 = \frac{1}{4} R_T K^2 \rho_{01} \rho_f [\alpha_f + (AA_v E_v/kT_0^2 \rho_{01}) \exp(-E_v/kT_0)] I_1^3 \tag{6.23}$$

The first term is proportional to the temperature coefficient of the film and is related to the scattering of the carriers by phonons, and the second term to the modulation of the vacancy concentration by the signal. This first term may be neglected when the vacancy concentration in the film is sufficiently high. In such a case, the third harmonic is proportional to the quasi-equilibrium vacancy concentration, which accounts for its exponential dependence on the temperature and the mechanical stresses, since $E_v = u_v - \sigma V_v$, so that:

$$U_3 = \frac{1}{4} R_T K^2 \rho_f (AA_v E_v/kT_0^2) I_1^3 \exp[-(u_v - \sigma V_v)/kT_0] \tag{6.24}$$

Figure 6.5 displays the experimental dependence of $U_3 I_1^{-3}$ on the mechanical stress, σ, using semi-logarithmic coordinates for Cr and Mo films fed by a small sinusoidal signal with a frequency $f_1 = 10\,\text{kHz}$ and a power $P \le 10\,\mu\text{W}$ [6.8]. These are the same film samples that were used to study the dependence of the PSD of the noise shown in Figure 5.15. The activation volumes obtained from these plots using the equation:

$$V_a = 2.3kT \frac{\Delta \log(U_3/I_1^3)}{\Delta \sigma} \tag{6.25}$$

are equal to $(1.3 \pm 0.4) \times 10^{-29}\,\text{m}^3$ and $(1.1 \pm 0.3) \times 10^{-29}\,\text{m}^3$ for Cr and Mo films respectively. These values coincide with the values of the activation volumes found from the dependence of the PSD of the $1/f$ noise on the internal mechanical stresses from Figure 5.16.

Computed and experimental values of U_3 for Mo films produced by the ion sputtering technique agree if one takes the atomic concentration of vacancies in equation 6.23 equal to $n_v \sim 10^{-3}\text{--}10^{-4}$, which is quite realistic for thin films.

Molybdenum films with an elevated content of reactive gas contaminants, produced by magnetron sputtering with a low rate of condensation, have a higher value of U_3 than expected from equation 6.23. This is explained by the presence of non-metallic (barrier) contributions to the resistance, which make an additional contribution to the third harmonic amplitude. This is confirmed by the low values of the temperature coefficient for these samples.

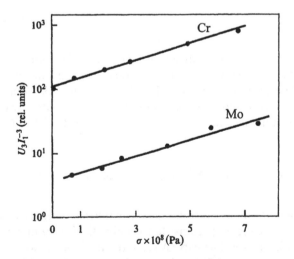

Figure 6.5 The measured current–voltage characteristic non-linearity coefficient against the internal mechanical stresses in Cr and Mo films [6.8]: $f_1 = 10\,\text{kHz}, P \le 10\,\mu\text{W}$.

To summarise, the calculation of the third harmonic amplitude and the experimental results confirm that a vacancy mechanism is responsible for both the $1/f$ noise and the cubic non-linearity of the current–voltage characteristic in metal films with an elevated concentration of mobile defects. This mechanism involves the creation and annihilation of vacancies at various types of sources and sinks. The coincidence of the activation volume values determined from these two measurements demonstrates that the vacancy mechanism is the common cause.

It is worth noting that the creation and annihilation of vacancies associated with the generation of quasi-equilibrium $1/f$ noise occurs at the expense of the internal energy of the crystal. The process of creation and annihilation of vacancies obeys the thermodynamic laws, since they are caused by local temperature fluctuations in the crystal lattice and involves defects with a variety of relaxation times, or activation energies.

The generation of the third harmonic of voltage is associated with the activation of micro-defects due to the Joule heating of the film by the weak signal at the fundamental frequency. A signal at a frequency f_1 changes the film temperature along with the power dissipation at $\sin^2 \omega_1 t$ or $2\omega_1$, which results in a vacancy concentration varying at this power frequency and the occurrence of non-linear effects. The defects with the lowest activation energy are generated most readily, which accounts for the coincidence of the activation energies for the third harmonic signal-response and the PSD of the $1/f$ noise at high frequencies ($f > 10^3 - 10^4$ Hz).

For films with a low vacancy concentration, the second term in equation 6.23 may be neglected and the amplitude of the third harmonic is given by:

$$U_3 = \frac{1}{4} R_T K^2 \rho_0 \rho_f \alpha_f I_1^3 \tag{6.26}$$

The temperature dependence is determined by the weak dependence of $\rho_f(T)$. Such films are characterised by a low amplitude of the third harmonic signal and a linear temperature dependence. The $1/f$ noise in these films is also weak. The noise level reaches a minimum and is estimated using equation 5.54 to have a value near the Hooge constant, $\alpha_H = 2 \times 10^{-3}$ [6.9].

6.6   Experimental results on the CVC quadratic non-linearity in metal films

It follows from the calculations in Section 6.5 that carrier scattering from phonons and equilibrium vacancies gives rise to the cubic CVC non-linearity but fails to produce quadratic non-linearity. A second harmonic, as well as a third harmonic has been reported [6.3] for the signal response from Al films which had been damaged by stressing with a large current and at a high temperature. The second harmonic was generated by the large Joule heating at local high resistance damage spots. The resistance change at $2f_1$ was revealed as a voltage signal at $2f_1$ by the large direct current.

A quadratic non-linearity was also observed in Mo films, prepared by magnetron sputtering, at a relatively low current density ($j < 0.5\times 10^4 \, \mathrm{A\,cm^{-2}}$) [6.1]. The quadratic non-linearity of metallic films may be caused by the resistance of barriers due to thin oxide layers at crystal boundaries (see Section 2.2.3). The U_2 versus U_1 dependence plotted in ($\log U_2, \sqrt{U_1}$) coordinates is linear for the films studied. This indicates that the non-linearity is caused by the emission of carriers over the barrier. If metal films with a high affinity for oxygen are formed at a low deposition rate, $w_c \approx 1 \, \mathrm{nm\,s^{-1}}$, thin ($\sim 0.1-1 \, \mathrm{nm}$ thick) metal oxide interlayers are formed at the grain boundaries because oxygen molecules are trapped by the condensate and some grains become insulated. In such films, the resistance of these barriers with conduction over the barrier results in a quadratic non-linearity and an increased contribution to the cubic non-linearity.

6.7 Non-equilibrium 1/f noise in metal and alloy films caused by excess vacancies

The voltage fluctuations across a film sample during the passage of a direct current are, according to the expansion 6.1:

$$\Delta U(t) = \Delta R_0(t)I_0 + \Delta R_1(t)I_0^2 + \Delta R_2(t)I_0^3 \tag{6.27}$$

where $\Delta U(t) = U(t) - \overline{U}$ and $\overline{U}$ is the time-averaged voltage drop across the sample. The fluctuations $\Delta R_0(t) = R_n(t) - \overline{R_n}$ ($n = 0, 1, 2$) of the $R_0(t)$ term give the equilibrium $1/f$ noise. $\overline{R_n}$ are the time-averaged coefficients. The fluctuations $\Delta R_1(t)$ and $\Delta R_2(t)$ are connected with the non-equilibrium conductivity fluctuations.

If the fluctuations of the coefficients $R_0(t)$, $R_1(t)$, $R_2(t)$ in (6.1) are not correlated, the spectral density of $U(t)$ is:

$$S_U(f) = S_{R0}(f)I_0^2 + S_{R1}(f)I_0^4 + S_{R2}(f)I_0^6 \tag{6.28}$$

where $S_{R0}(f)$, $S_{R1}(f)$, and $S_{R2}(f)$ are the spectral densities of the fluctuations of the coefficients R_0, R_1, and R_2, which can be calculated using equations 6.8–6.10 from the measured PSDs of the amplitude fluctuations of the harmonics U_0, U_2, and U_3.

If the fluctuations of the coefficients are correlated, then the third- and fifth-order terms also contribute to the current dependence of $S_U(f)$. Thus, non-equilibrium conductivity fluctuations result in deviations of the current dependence of the power spectral density of the $1/f^\gamma$ noise from a quadratic dependence. These deviations should be largest at high currents.

Non-equilibrium $1/f$ noise measurements made during the passage of a direct current through the sample were carried out on two types of Cr films with high concentrations of mobile defects (mainly vacancies) (Type A) and fixed defects (Type B) and also on Al/Cu alloy films with a low concentration of mobile defects [6.10]. For the latter fresh (Type C) samples and samples damaged by electromigration stress (Type D) were investigated. The Cr films of Type A were deposited on glass substrates by thermal evaporation in a vacuum. The Cr films of Type B were deposited on oxidised silicon wafers by thermal evaporation in an atmosphere of nitrogen. The Al/Cu alloys films were deposited on oxidised silicon wafers by magnetron sputtering. All the samples had metallic conductivity [6.10].

For the Cr films of Type A the experimental dependence of the PSD of the voltage fluctuations on the direct current (Figure 6.6) fits well to a power law $S_U \sim I_0^a$ where the exponent a changes with current. It is equal to 2.0 for low current densities, $j_0 < 5 \times 10^3$ A cm^{-2}($I_0 < 2$ mA) but reaches 4.0 for $j_0 > 10^4$ A cm^{-2}($I_0 > 4$ mA). Similar dependencies have also been observed for Ni–Cr strips [6.11] and Al films [6.12].

The quadratic dependence of S_U on the current for the Cr films of Type 1 is the equilibrium $1/f^\gamma$ noise where the voltage fluctuations are due to the fluctuations of the current-independent part, $R_0(t)$, of the film resistance (equation 6.1). For the particular Cr films studied the equilibrium noise is caused by fluctuations in the number of quasi-equilibrium vacancies. The frequency exponent is close to 1.0 over the frequency range 2–10 kHz. The vacancy mechanism of the noise is confirmed by the temperature dependence of the PSD which shows an activation energy $E_v \approx 0.6$ eV at a frequency of 35 Hz. The noise level is high and characterised by the parameter (equation 5.59) $\alpha = 10^1 - 10^2$, where the free carrier concentration is taken as $n_0 = 1.7 \times 10^{22}$ cm^{-3}. The noise is a resistance fluctuation due to changes in the number of vacancies, $N_v(t)$. The contribution to the film resistivity is given by equation 5.60.

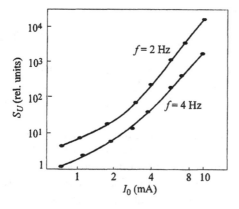

Figure 6.6 The PSD of the voltage noise, S_U, of a Cr film Type A against the direct current stress [6.10].

For the vacancy mechanism, equation 5.66, the noise PSD is:

$$S_U = KI_0^2 \overline{n_v^2}/f = KI_0^2 n_v/N_a f \tag{6.29}$$

where K is a constant for the film sample, $n_v = N_v/N_a$ is the average atomic concentration of quasi-equilibrium vacancies and $\overline{n_v^2}$ is the variance of n_v. Here it is assumed that the variance of the quasi-equilibrium vacancy number fluctuations in the sample, $\overline{N_v^2}$, is equal to their mean number N_v. At room temperature, T_0, we have from equation 4.5 a vacancy concentration:

$$n_{v0} = A_v \exp[-(E_v/kT_0)] \tag{6.30}$$

When a direct current is applied, the film temperature rises because of Joule heating by an amount:

$$\Delta T = T - T_0 = R_T R I_0^2 \tag{6.31}$$

where R_T is the thermal resistance and R is the film specimen resistance. The vacancy atomic concentration increases because of the film heating and is given by:

$$n_v = A_v \exp[-E_v/k(T_0 + \Delta T)] \tag{6.32}$$

For $\Delta T \ll T_0$, we find that:

$$n_v \approx n_{v0} \exp(E_v \Delta T/kT_0^2) \tag{6.33}$$

Substitution of equation 6.31 into equation 6.33 and expanding in powers of ΔT, assuming that $E_v \Delta T/kT_0^2 \ll 1$, yields an equation for the vacancy concentration under Joule heating, taking only the first three terms:

$$n_v \equiv n_{v0}(1 + BI_0^2 + \frac{1}{2}B^2 I_0^4) \tag{6.34}$$

where $B = E_v R_T R/kT_0^2$ is a constant.

Taking into account that $\overline{n_{v0}^2} = n_{v0}/N_a$, the equation for the variance of the vacancy concentration, we obtain:

$$\overline{n_v^2} \approx (n_{v0}/N_a)(1 + 2BI_0^2 + 2BI_0^4) \tag{6.35}$$

Equations 6.29 and 6.35 define the dependence of the PSD of the $1/f^\gamma$ noise on current for the vacancy mechanism:

$$S_U = KI_0^2(n_{v0}/N_a)(1 + 2BI_0^2 + 2BI_0^4)/f \tag{6.36}$$

The first term in equation 6.36 gives the equilibrium $1/f^\gamma$ noise. Other terms are the non-equilibrium $1/f^\gamma$ noise. The second term gives the dependence of $S_U \sim I_0^4$. For currents $I_0 > (2B)^{-1/2}$ this term is greater than the first one and the non-equilibrium $1/f^\gamma$ noise is larger than the equilibrium noise. At the current value $I_0 = I_m = (2B)^{-1/2}$ the concentration of the additional vacancies generated by the Joule heating is equal to the equilibrium concentration. For the sample of Type A the numerical data are $E_v = 0.6\,\text{V}$, $R_T R = 4 \times 10^5\,\Omega\,\text{K}\,\text{W}^{-1}$ giving $I_m = 4\,\text{mA}$ and the current density $j_m \approx 7 \times 10^3\,\text{A}\,\text{cm}^{-2}$. At this current one may thus expect the change of the slope seen in Figure 6.6.

The experimental dependence of the PSD of the $1/f^\gamma$ noise for a Cr film of Type B against the direct current density squared and the values calculated using the Hooge formula (equation 5.54) with $a_H = 10^{-3}$ are given in Figure 5.22. The feature of Cr films with a high concentration of fixed defects and few mobile ones (Type B) is that they show a low noise level with a Hooge parameter value $a_H = 10^{-3}\text{–}10^{-5}$ at a current density $j_0 < 10^6\,\text{A}\,\text{cm}^{-2}$. In films of Type B the $1/f$ noise at a current density $j_0 < 10^6\,\text{A}\,\text{cm}^{-2}$ is caused by fluctuations in the mobility because of lattice scattering. It is very small because the fixed defects decrease the mobility related fluctuations. At sufficiently high current density non-equilibrium noise appears because of the generation of additional mobile defects when the film specimen heats up. For the non-equilibrium $1/f^\gamma$ noise the PSD increases as $S_U \sim j_0^4$ or j_0^6 in accordance with equation 6.36. This takes place, if $2BI_0^2 \gg 1$, since for these films the first term $KI_0^2(n_{v0}/N_a)$ in equation 6.36 does not contribute to the $1/f$ noise at all.

For the undamaged samples of the Al/Cu alloy (Type C) the non-equilibrium $1/f$ noise is not observed at low $(T = 297\,\text{K})$ or high $(T = 523\,\text{K})$ stress temperatures for current densities $j_0 \leq 2.2 \times 10^6\,\text{A}\,\text{cm}^{-2}$. The noise shows an ordinary j_0^2 dependence for S_U as seen by the open circles in Figure 6.7. This

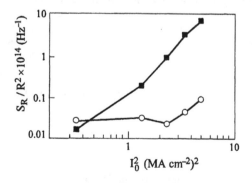

Figure 6.7 The relative resistance fluctuations against DC current squared [6.10]. For the undamaged sample (open circles, Type C) the value is constant implying $S_U \sim I^2$ while for the damaged sample (solid squares, Type D) the values vary as I^2 implying that $S_U \sim I^4$.

is because the concentration of additional generated mobile defects is very small. The value of the Hooge parameter (equation 5.59) is found to be $\alpha = 8 \times 10^{-3}$ with the assumption that $n_c = 1.8 \times 10^{23}\,\mathrm{cm}^{-3}$.

In the damaged samples (Type D) the non-equilibrium noise with $S_U \sim j_0^4$ as given by the solid squares in Figure 6.7. This may be explained by the appearance of local self-heating or areas of high current density in the film due to the non-uniform distribution of the electromigration damage along the line. There can also be a contribution due to the creation and healing of voids.

These results demonstrate that stationary non-equilibrium $1/f^\gamma$ noise in metal films can occur because of the generation of additional vacancies by Joule heating. For the Al films the non-equilibrium noise may precede the electromigration processes and serve to indicate the early stages of electromigration degradation.

6.8   The measurement of non-equilibrium $1/f$ noise

According to equation 6.28, fluctuations of all the terms in the series (equation 6.1) contribute simultaneously to the measured power spectral density of the $1/f$ noise at a given direct current. This prevents us from determining the separate contribution of each term to the $1/f$ noise. At small direct currents the non-equilibrium noise may also be masked by equilibrium noise, as happens for the Cr films at $I_0 < 2\,\mathrm{mA}$ (Figure 6.6). At high direct current the non-equilibrium noise in metal films due to Joule heating may also mask the non-equilibrium noise due to structural heterogeneity. So other measurement methods are needed to separate the non-equilibrium noise from the total noise PSD.

The experimental method is based on measuring the spectra of the amplitude fluctuations for the third, $U_3(t)$, second $U_2(t)$ and zero $U_0(t)$ harmonics of the voltage produced by a sinusoidal current applied to the sample.

Spectral measurements of the amplitude fluctuations of the various harmonics (equations 6.8–6.10) can be used to determine the PSDs of the coefficients in the expansion (equation 6.1) of the resistance. This allows the non-equilibrium $1/f$ noise to be separated from the total noise PSD and makes it possible to study non-equilibrium conductivity fluctuations.

A special device is used to measure the PSD of the amplitude fluctuations for harmonic components of the film response to sinusoidal excitations [6.1, 6.13]. The voltage responses for the third $U_3(t)$ and second $U_2(t)$ harmonics are extracted with the help of the circuit for the measurement of the non-linearity of the CVC (Figure 6.1), amplified by the selective microvoltmeter, detected and measured by the spectrum analyser. The test signal frequency in these experiments was $f_1 = 10\,\mathrm{kHz}$. The PSD of the amplitude fluctuations near the sinusoidal test signal ($1/\Delta f$ noise), zero, third and second harmonics were measured in the frequency range from $10\,\mathrm{Hz}$ to $1\,\mathrm{kHz}$. The experimental results of the non-equilibrium noise in molybdenum films are reported in the next section.

6.9   Experimental results on non-equilibrium $1/f$ noise in molybdenum films

As well as the spectra of the amplitude fluctuations for the third $U_3(t)$, second $U_2(t)$ and zero $U_0(t)$ harmonics, the PSD of the excess, $1/f$, noise was measured when a direct current was passed through the sample. The direct and alternating current densities were less than $10^4\,\mathrm{A\,cm^{-2}}$ to rule out non-equilibrium fluctuations caused by Joule heating and local self-heating.

Molybdenum films on oxidised silicon were prepared by magnetron sputtering at condensation rates of 1 and $2.4\,\mathrm{nm\,s^{-1}}$. The films deposited at the lower rate had increased impurity concentrations of reactive gases, mainly oxygen. For Mo films 247 and 560 nm thick, both 40 µm wide, and deposited at rates of 1 and $2.4\,\mathrm{nm\,s^{-1}}$, respectively, the resistivity of the films 247 nm thick $\rho_f = 34\,\mu\Omega$ cm; the temperature resistance coefficient $\alpha_f = 0.55 \times 10^{-3}\,\mathrm{K^{-1}}$; and for films 560 nm thick, $\rho_f = 11\,\mu\Omega$ cm, $\alpha_f = 2.6 \times 10^{-3}\,\mathrm{K^{-1}}$.

Excess low frequency noise for the Mo films deposited at $2.4\,\mathrm{nm\,s^{-1}}$ was observed only at high current density when a direct current was passed through the sample. The noise level was below the thermal noise of the film specimen at a frequency of 1 kHz and a current density of less than $10^4\,\mathrm{A\,cm^{-2}}$. At higher current densities the films exhibited normal $1/f^\gamma$ noise with an exponent $\gamma \approx 1$ and an intensity much higher than predicted by Hooge's formula (equation 5.54). For these films the temperature dependence of the PSD of the $1/f$ noise was exponential and the high noise level is explained by an elevated concentration of mobile vacancies. The amplitude fluctuations of the zero, second, and third harmonics for this film at an alternating current density $j_1 = 10^4\,\mathrm{A\,cm^{-2}}$ were also small and were below the measurement sensitivity.

The films obtained at a low deposition rate, $w_c \approx 1\,\mathrm{nm\,s^{-1}}$, showed a higher $1/f$ noise level and non-linearity than those deposited at a rate of $2.4\,\mathrm{nm\,s^{-1}}$. For these films, the amplitudes of the second and third harmonics were approximately two and four orders of magnitude, respectively, higher, for $I_1 = 0.45\,\mathrm{mA}$.

Figure 6.8 shows the frequency dependence of the PSD of the excess noise at a direct current $I_0 = 0.45\,\mathrm{mA}$ (curve 1) and the noise spectral density of the side bands for the zero, second and third harmonics at an alternating current $I_1 = 0.45\,\mathrm{mA}$ (curves 2, 3, 4) for these films. It can be seen from the figure that the spectral density of the amplitude fluctuations for the zero harmonic coincide with one for the second harmonic within the limits of the measurement accuracy. This result agrees with equations 6.8 and 6.9.

The spectral density of the amplitude fluctuations of the third harmonic (curve 4) is less than the spectral density of the excess noise measured during the passage of a direct current through the sample (curve 1) by almost three orders of magnitude. The exponent $\gamma = 1.8 \pm 0.2$ for all the curves.

From these experimental results the relative contribution to the excess noise spectral density for direct current excitation was determined for both

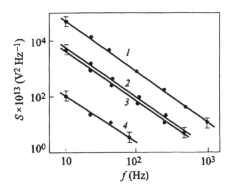

Figure 6.8 The power spectrum of the excess noise for direct current excitation $I_0 = 0.45$ mA (curve 1) and for the harmonic fluctuations under alternating current excitation $I_1 = 0.45$ mA for the zero, second and third harmonics (curves 2, 3, 4) respectively [6.13].

the linear and non-linear terms of the resistance fluctuations in accordance with equation 6.28. The relative contributions of each term in equation 6.28 to the PSD of the noise at 10 Hz and $I_0 = 0.45$ mA are 39.5, 60, and 0.5% for the first, second, and third terms, respectively. The largest contribution to the flicker noise was the fluctuation of the quadratic term. The contribution of the $R_1 I_1$ term to the film resistance (equation 6.1) at $I_1 = 0.45$ mA was less than 0.1%.

Such a large contribution by the $\Delta R_1 I_0^2$ term in equation 6.27 to the excess noise level while there is negligible contribution of $R_1 I_1$ term to the film resistance is explained by the high level of the fluctuation in the coefficient $\Delta R_1(t)$ in equation 6.27 due to the non-metallic conductivity displayed in metal films, deposited at a low rate.

Figure 6.9 displays the frequency dependence of the PSD of the amplitude modulation, measured for alternating current $I_1 = 0.45$ mA, for the second m_1 (curve 1) and third m_2 (curve 2) harmonics and of the relative PSD of flicker noise $S_U(f)/U^2$ measured for a direct current $I_0 = 0.45$ mA (curve 3).

The degree of noise-induced amplitude modulation of the second harmonic (term R_1) is six orders of magnitude higher than the level of relative fluctuations of the voltage applied to the sample.

The quadratic non-linearity of metal films may be caused by Schottky field emission through the thin oxide layers at the crystal boundaries (see Section 2.2.3). The fluctuations of R_1 in equation 6.1 are caused by fluctuations of the potential barrier heights at grain boundaries that generate non-equilibrium conductivity fluctuations.

In high-quality metal films with a low concentration of defects, where the resistivity is close to that of bulk metals, equilibrium $1/f$ noise is observed with a level given by the Hooge formula (equation 5.54) with the constant $\alpha \approx 2 \times 10^{-3}$.

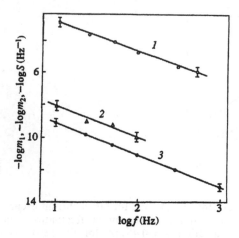

Figure 6.9 The frequency dependence of the spectral density of the fluctuations of the amplitude modulations for the second (curve 1) and third (curve 2) harmonics of an alternating current $I_1 = 0.45\,\text{mA}$ and the relative spectral density of the flicker noise intensity (curve 3) for a direct current $I_0 = 0.45\,\text{mA}$ [6.13].

To summarise, the $1/f$ noise in metal films with a high concentration of impurities may be generated by fluctuations of both the linear and non-linear CVC terms, and the contribution of the latter may be dominant. In each case the level of the fluctuations is determined by the scattering of charge carriers.

Literature for Chapter 6

6.1   Zhigal'skii, G.P., "Nonequilibrium flicker noise in conducting films", *Russian J. Phys. Chem.*, **69** (1995) 1218–1220.

6.2   Kirby, P.L., "The non-linearity of fixed resistors", *Electronic Engineering*, **37** (1965) 722–726.

6.3   Jones, B.K. and Xu, Y.Z., "Characterisation of electromigration damage by multiple electrical measurements", *Microelectron. Reliab.*, **33** (1993) 1829–1840.

6.4   Jones, B.K., "$1/f$ and $1/\Delta f$ noise produced by a radio-frequency current in a carbon resistors", *Electron. Lett.*, **12** (1976) 110–111.

6.5   Dobrynina, G.M., Zhigal'skii, G.P. and Mozgin, A.A., "Investigation of CVC non-linearity in chromium films", *Fizika Poluprovodnikov i Mikro-elektronika (Physics of Semiconductors and Microelectronics)* (in Russian) RRTI Publ., 1979, 79–82.

6.6   Zhigal'skii, G.P., Kurov, G.A. and Siranashvili, I.Sh., "Excess noise and mechanical stress in thin chromium films", *Radiophys. Quantum Electron.*, **26** (1983) 162–166.

6.7   Zhigal'skii, G.P., Markaryants, E.A. and Fedorov, A.S., "Investigation of electro-physical properties of the refractory metal films manufactured by ionic sputtering", *Poverkhnost': Fizika, Khimiya, Mekhanika.*, 1993, 78–86.

6.8   Zhigal'skii, G.P., "Relationship between $1/f$ noise and non-linearity effects in metal films", *JETP Lett.*, **54** (1991) 513–516.

6.9   Zhigal'skii, G.P., "$1/f$ noise and non-linear effects interaction in metal films", *Proc. Int. Conf. on Noise in Physical Systems and $1/f$ Fluctuations*, Musha, T., Sato, S. and Yamamoto, M. (eds), 1991, Kyoto, 39–42.

6.10   Zhigal'skii, G.P. and Jones, B.K., "Non-equilibrium $1/f$ noise in metal and alloy films", *Proc. Int. Conf. on Noise in Physical Systems and $1/f$ Fluctuations*, Surya, C. (ed.), 1999, Hong Kong, 172–175.

6.11   Touboul, A., Verdier, F. and Herve, Y., "Lf noise measurements applied to the determination of electromigration mechanisms in Al (Si,Ti) and Ni-Cr stripes", *Proc. Int. Conf. on Noise in Physical Systems and $1/f$ Fluctuations*, 1991, Kyoto, 73–77.

6.12   Schwarz, J.A., Patrinos, A.J., Bakshee, I.S., Salkov, E.A. and Khizhnyak, B.I., "Grain size dependence of $1/f$ noise in Al-Cu thin-film interconnections", *J. Appl. Phys.*, **70** (1991) 1561–1564.

6.13   Zhigal'skii, G.P., "Investigation of nonequilibrium conductivity fluctuation in metal films", *Proc. 7th Vilnius Conf. Fluctuation Phenomena in Physical Systems*, 1994, Palanga Palenskis, V. (ed.), Vilnius University Press, 61–64.

7 Degradation mechanisms in thin metal films

The interconnections between the transistors within an integrated circuit (IC) are subjected to current and thermal stresses during operation. This creates favourable conditions for the operation of various degradation processes which produce a gradual modification in the electrical properties of the conducting films, and then their sudden failure. An analysis of the types of failure in modern ICs shows that about 25–30% of failures of the chips can be accounted for by the metal interconnections [7.1]. This indicates the need to study the degradation mechanisms in thin metal films. As well as the catastrophic failures due to major electrical overstress, the basic mechanisms of degradation are electromigration and corrosion [7.1–7.5].

7.1 Electromigration in thin films

Electromigration is the transport of mass in metals when the metals are stressed at high current densities. The phenomenon of electromigration appears in metal conductors for a direct current density $j \geq 10^6 \, \mathrm{A\,cm^{-2}}$. The mass transport of the conductor material takes place from the region of the negative contact (cathode) to the positive contact (anode). Thus the electromigration process is a transfer of atoms, both the host and impurity, caused by the passage of the direct current. This effect has been known for more than five decades and has been observed in both molten and solid metals.

The basic mechanism is the following [7.5]. An electric field produces a flow of electrons in a direction opposite to the electric field. The positively charged ions of the metal move under the action of two forces. One is an electrostatic force, F_z, which acts along the field and tends to transfer atoms in the direction of the cathode. The second force, F_e, is due to the interaction of the electron flow with the ions of the metal, because of a frictional force between the current carriers and the atoms of the lattice. Because of this force the atoms in the conductor gain an additional momentum in the direction of motion of the electron flow, the anode direction. At large enough current density, $F_e > F_z$, the atoms of the metal will move from the end of the metal conductor with a negative potential towards the positive potential contact. As a result voids are created in the region of the negative potential and accumulation occurs at the positive potential. In some areas, film growths called whiskers or hillocks will form. If there is a gradient of temperature along a conductor the process of transportation may be accelerated.

Note that for an alternating current little net electromigration occurs because the alternating motions of the atoms in each half cycle act in opposite directions so that the electromigration effects compensate if the process does not produce an irreversible effect on each cycle. Electromigration in conductors can also be neglected for direct currents of low density, $j < 10^4\,\mathrm{A\,cm^{-2}}$, which is the usual case for bulk conductors, and at temperatures very much lower than the melting temperature of the metal.

7.1.1 Failure modes caused by electromigration

Electromigration in thin film conductors produces two types of failures. First there is a significant decrease in the cross-section of the conductor at a point due to the formation and joining together of voids to result in the rupture of the film when the sizes of the resulting voids is comparable with the cross-section of the film. Note that this is a cumulative effect since the reduction in the cross-section produced by the creation of a void produces an increase in the current density, the local heating and the temperature with increased electromigration. The result can then be very rapid.

The second type of failure is caused by the accumulation of the metal in some kind of hillock or whisker in an area of the film resulting in a short circuit to an adjacent interconnect line or a track located at a different level of a multilayer interconnection. Also the accumulation of metal may cause a rupture of the protective dielectric capping and result in a later failure by corrosion (see Section 7.2).

7.1.2 Median time to failure of thin-film conductors

The measurement of the median time to failure of thin-film conductors is the usual method to evaluate their resistance to electromigration failure. Research on electromigration in thin films is usually carried out on groups of a statistically significant number of identical conductors during accelerated tests at high current densities ($j \approx 10^6 - 10^7\,\mathrm{A\,cm^{-2}}$) and at elevated temperature ($T \approx 100 - 500\,^{\circ}\mathrm{C}$). From these tests the median time to failure is determined as the time for which half of the conductors from the total number have failed by open circuit [7.2, 7.5].

To predict the failure rate at low temperatures and small current densities an empirical relationship for the failures during accelerated tests was developed, in which a simplified relation between the median time to failure,

t_{50}, current density, j, and the film conductor temperature is given by the Black expression [7.2, 7.5]:

$$t_{50} = A j^{-n} \exp(E_a/kT) \qquad (7.1)$$

where E_a is the activation energy for the electromigration process, which can be derived from a fit of this equation to accelerated test data at several temperatures and current densities or obtained from the measurement of resistance change during stress tests; k is the Boltzmann constant; T is the absolute temperature; the constant of proportionality, A is the performance of the conductor, which depends on the film material, its structure, length, breadth and thickness. The current exponent, n, derived from experiment is found to be unity at small current densities ($j < 10^5 \, \mathrm{A \, cm^{-2}}$) and increases at larger current densities, to values in the range from 1 to 10 and even higher as shown in Figure 7.1 [7.2, 7.5]. Here three curves are presented, which correspond to different conditions of heat dissipation. They were obtained with an ambient temperature of 200 °C and the initial temperatures of the conductor were 205, 210 and 220 °C since there is an increase in temperature ΔT_R of the film temperature over the ambient due to Joule heating. These

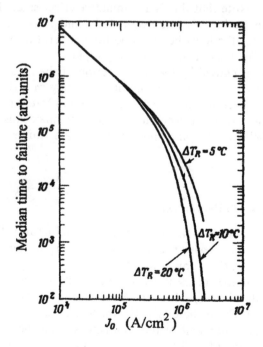

Figure 7.1 Dependence of the median time to failure on the current density [7.2]. The curves correspond to different conditions of heat dissipation. ΔT_R is the increase of the film temperature over the ambient temperature due to Joule heating.

cases are close to the real working conditions, since there is always a thermal resistance between the film and substrate so that some self-heating inevitably appears.

It should be noted that the theory of electromigration for bulk metals gives the value $n = 1$, which corresponds to experiment. The increase in the index n with current density for film conductors can be explained by the structural singularities of film and the temperature gradients in the film. In the case when the dominating factor for the increase in failure rate is the inhomogeneous heating of the thin-film conductor, the value $n = 3$. Often for thin metal films at $10^5 \le j \le 2 \times 10^6 \, A \, cm^{-2}$ the index n is taken to be 2.

The distribution of time to failure due to electromigration at a constant stress is usually close to a logarithmic normal distribution. Thus the probability of failure during time t is given by the expression [7.2, 7.5]:

$$p = \frac{1}{\sqrt{2\pi}\sigma_f t} \exp\left[-\frac{1}{2}\left(\frac{\ln(t/t_{50})}{\sigma_f}\right)^2\right] \tag{7.2}$$

where σ_f is the standard deviation characterising the breadth of the distribution and depends on the number and type of failures.

7.1.3 Basic theory of electromigration in conductors

The basic mechanism of the atom motion during electromigration is diffusion. The random diffusion motion is changed to a net directed motion by the interaction of the conduction electrons with the atoms of the metal. This interaction is called the "electron wind". This is essentially the force of friction between the ions of the metal and the flow of electrons due to the electrical resistance processes. This mass transport can be of both the lattice or impurity ions.

If there is a potential difference in a conducting material then there is a directed flow of the charge carriers, electrons or holes. In the path of the moving electron there can be a thermally excited atom, which has obtained a sufficiently large amount of thermal energy accidentally to make an elementary diffusion jump. The excited ion is on top of the potential barrier which divides two adjacent positions of equilibrium (Figure 7.2.) and is subjected to the operation of the electron wind. By its motion in an electrical field the electron can transmit to the ion a part of its momentum. The force of the electron wind, F_e, is directed in the direction of the electron flow. In the free electron approximation the force of the electron wind is [7.5]:

$$F_e = -n\sigma_{ei}leE = -Z_{ei}eE \tag{7.3}$$

where n is the concentration of the conductivity electrons, e is the electron charge (absolute value), l is the electron mean free path and σ_{ei} is an average

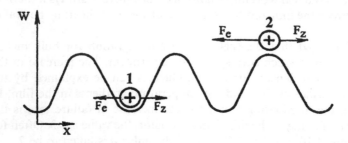

Figure 7.2 Forces acting on the normal (1) and thermally activated (2) ions. W is the potential energy of the ion. F_e is the force of the electron wind and F_z is the electrostatic force. There is net resulting force acting on the activated ion.

over all angles of the cross-section of scattering of electrons by activated ions. The values of σ_{ei} and l correspond to those of the conduction electrons with an energy equal to the Fermi energy and can be evaluated from the resistivity. The effective charge of the ions is given by:

$$Z_{ei} = n\sigma_{ei}l \tag{7.4}$$

Note that the product leE in equation 7.3 is the energy gained by an electron from the electrical field in moving a distance equal to the mean free path. The electric field applied to the conductor will exert an electrostatic force F_z on the activated positively charged ion in a direction opposite to the electron flow. This force is directed to the cathode and equal to the product of the ion charge $q_i = Ze$ and the electric field and can be written as:

$$F_z = q_iE = ZeE \tag{7.5}$$

where Z is the valence of the atom (the number of conduction electrons on one atom). The force F_z is directed opposite to the electron flow. The resulting force F_i is equal to the sum of the forces F_e and F_z so that from equations 7.3 and 7.5 the full force is:

$$F_i = (q_i - n\sigma_{ei}le)E = q_{ef}E \tag{7.6}$$

where q_{ef} is an effective charge of the activated ion in the electrical field.

$$q_{ef} = q_i - n\sigma_{ei}le = Z^*e \tag{7.7}$$

Here Z^* is an effective charge for the process expressed in terms of the electron charge (absolute value):

$$Z^* = Z - n\sigma_{ei}le \qquad (7.8)$$

From equation 7.6 it follows that the direction of the force and the atom motion are determined by the sign of the effective charge and the relative size of the two contributions in equation 7.7. If $q_{ef} > 0$ the resulting force is directed with the field and the atoms will move in the direction of the cathode. If $q_{ef} < 0$ the resulting force is directed opposite to the electric field and atoms will move towards the anode and the force of the electron wind dominates. The force of the electron wind in metals is much greater than the electrostatic force F_z, which depends on the true charge of the ion, Ze, but is screened by the carriers. Using equations 7.3 and 7.5 we obtain the ratio of these forces as:

$$\alpha = \frac{|F_e|}{F_z} = \frac{n\sigma_{ei}l}{Z} \qquad (7.9)$$

Substituting some typical values for metals: $n = 5 \times 10^{22}\,\mathrm{cm}^{-3}$, $\sigma_{ei} = 10^{-15}\,\mathrm{cm}^{-2}$, $l = 10^{-6}\,\mathrm{cm}$ into equation 7.9 and assuming that $Z = 1$, we obtain $\alpha = 50$. That is the force of the electron wind is 50 times more than the force of electrostatic interaction. Thus the effective charge of an ion is, according to equations 7.7 and 7.8, $Z^* = -49$. Since this effective charge is negative the motion of positively charged ions to the anode is the result of the electron wind.

The above is for the motion of the lattice atoms due to the phonon contribution to the resistance. An impurity ion in a metal can differ from the host ion by its charge and scattering cross-section, $\sigma_{ei} \neq \sigma_0$. Thus the effective charge of the activated atoms in a pure metal is not equal to zero but, by scaling equation 7.7 is:

$$q_{ef} = Ze\left(1 - \frac{\sigma_{ei}}{\sigma_0}\right) \qquad (7.10)$$

Experiments have shown that the transport of ions by electromigration occurs along the boundaries of grains (grain boundary diffusion) in the same way as within a crystal lattice. Electromigration along the grain boundaries is caused by the motion of atoms on the grain boundaries between crystallites. It is the prefered path with a lower activation energy than the bulk since the lattice is disrupted with a lower average density than the bulk. Other paths are along dislocation cores and interfaces with the oxide or vacuum. Each of the distinct mechanisms of mass transport is characterised by its diffusion coefficient, activation energy and effective charge.

7.1.4   Electromigration in thin films

Free wires of pure metal are warmed up and even melted at a current density of $j \approx 10^4$ cm^{-2}. Thin metal films on substrates can survive at higher current densities, up to 10^6 A cm^{-2}, with only a moderate increase of temperature $\Delta T \approx 10$–20 °C due to Joule heating. This is because of the very effective cooling of a film by conduction to the substrate along its wide surface. With an increase in the degree of integration of ICs the cross-sectional area of the interconnects decreases so that this is accompanied by a corresponding increase in the current density. Thus electromigration as a potential failure mode for ICs becomes a serious obstacle to the creation of reliable thin-film conductors with sub-micrometer sizes. For example, for aluminium interconnect lines, as the width of the tracks decreased from 10–15 μm to 1–2 μm the median time to failure decreased by 10 and more times [7.2]. Other scaling assumptions for the components of ICs result in similar conclusions about the growing importance of electromigration failure as the integration becomes more dense.

At high current density and high operating temperatures the electromigration phenomena in films becomes more serious. Thus the mechanism of electromigration in thin films differs essentially from the mechanism of electromigration in bulk materials. In bulk samples the basic mechanism of transport of matter is the diffusion in the crystalline lattice (bulk lattice diffusion). In films with a polycrystalline structure, at fairly low temperatures (up to +200 °C) mass transport occurs mainly along the grain boundaries since the activation energy of diffusion along the grain boundaries is much lower than in the bulk. For Al the activation energies are 0.5–0.7 eV and 1.3–1.4 eV. This is because the energy for atoms to jump from vacancy to vacancy in grain boundary diffusion is less than for bulk crystal lattice diffusion. Also the grain boundaries appear to be the preferred areas for the localisation of mobile imperfections and especially vacancies.

The diffusion along grain boundaries in polycrystalline materials takes place tens of thousands of times faster than within the crystal lattice. It was found for electromigration in thin films of pure Al at $T = 175$ °C that the flow of the host atoms because of grain boundary diffusion, J_{gb}, is approximately 10^6 times higher than the flow in a crystal lattice, J_i. For the electromigration of impurity atoms of copper in aluminium films at $T = 225$ °C the ratio is 10^4 [7.2]. Measurements on single crystalline aluminium films have shown an almost complete reduction in the degradation over a long time. This confirms that electromigration in thin polycrystalline films occurs mainly along the grain boundaries.

In thin films, as well as grain boundary diffusion, it is necessary to take into account the other paths for mass transport: along dislocations and on the surface or interface of the conductor. It is known that surface diffusion has a lower activation energy than the bulk. Surface hollows are seen to develop under the effect of an electric current. In films the surface to volume ratio is much larger than in bulk conductors so that mass transport along the surface of the films can dominate.

In the analysis of the mass transport in thin films, it is necessary to take into account also the size effect in the conductivity that produces the dependence of the film properties on the thickness. The size effect in electromigration in thin films produces an increase in the effective ion charge q_{ef} and a decrease in the activation energy [7.2]. This in turn leads to an increase in the drift velocity of the ions and a decrease in the average time of operation before the destruction of a film.

Also, diffusion in thin films can arise from a force produced by a mechanical stress gradient in a film. This is observed in ICs primarily at steps in the oxide, SiO_2. Applying a mechanical force to a metal film from the back side of a substrate can enhance the process of electromigration. This also means that the mechanical stresses in a film are distributed inhomogeneously on the grain boundaries [7.2]. The calculation of all the paths of mass transport in thin films with such variety is a very difficult problem.

7.1.5 Mass transport in the crystal lattice

Let us derive an expression for the flow of atoms on a crystal lattice during electromigration. For this purpose we will calculate the drift velocity of atoms in the field of the external forces.

Let us use the Einstein relation between the diffusion coefficient and mobility, which is true for any charged particles diffusing in an external electric field. According to this relation the drift mobility of an ion in a metal is:

$$\mu_i = \frac{Dq_{ef}}{kT} \tag{7.11}$$

where D is the self-diffusion coefficient at the temperature T, and the drift velocity of an ion is expressed by means of the relationship:

$$\nu_i = \mu_i E = \frac{D}{kT} q_{ef} E \tag{7.12}$$

Using equations 7.7 and 7.12 we obtain:

$$\nu_i = \frac{D}{kT}(q_i - n\sigma_{ei}le)E = \frac{D}{kT}Z^*eE \tag{7.13}$$

That is the drift velocity of ions is proportional to the self-diffusion coefficient and the strength of the external electrical field. As the effective charge of an ion is negative, the velocity of an ion in equation 7.13 is directed against the field and towards the end of the conductor having the positive potential. Ions in a conductor gain additional momentum in the direction of the electrical flow. For the atom flow on a crystal lattice it is possible to write the flow of atoms through a unit square of the film cross-section in unit time [7.2, 7.5] as:

$$J_i = N_i \nu_i = N_i \frac{D_i}{kT} Z_i^* eE \qquad (7.14)$$

Here N_i is the concentration of diffusing ions (the density of mobile imperfections) and D_i is the self-diffusion coefficient. The index i indicates the directions of the crystal lattice.

7.1.6   Mass transportation along grain boundaries

Let us now write an expression for the flow of atoms along grain boundaries similar to equation 7.14. For simplicity we will consider a film with an ideal structure of hexagonal columnar grains (Figure 7.3) with a size equal to the thickness of the film and perpendicular to the plane of the film. For a film with such a structure the expression is [7.5, 7.6]:

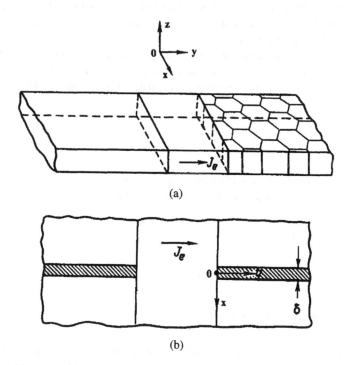

(a)

(b)

Figure 7.3 Geometric model of a film with an ideal hexagonal texture structure of the grains: (a) arrangement of the grains in the film; (b) schematic representation of the grain boundary [7.6].

$$J_{gb} = N_{gb} \frac{\delta \, D_{gb}}{d \, kT} Z_{gb}^* eE \qquad (7.15)$$

where N_{gb} is the concentration of the diffusing, activated, atoms in the grain boundaries, δ is an effective width of the grains boundary, which is usually taken as equal to 0.5–1 nm and d is the average size of the grains. The parameters describing the grain boundary diffusion are marked by the index "gb".

Equation 7.15 is obtained with the assumption that the transportation takes place on all the grain boundary paths with identical properties. For real films there is always a variation of the transport parameters on the separate grain boundaries produced by the random geometric orientation of the grain boundaries and their own structural properties. Therefore the accuracy of equation 7.15 for the flow of atoms on a grain boundary is lower than that of formula 7.14 for the flow on the bulk lattice. Also the transport parameters are not well known for grain boundaries. As the exact structure of the grain boundaries is unknown, it is impossible to indicate exact values for parameters N_{gb} and Z_{gb}^*. For simplicity we will assume that the parameters of the grain boundary in expression 7.15 are averaged to allow for the random orientation of the grains.

If we bring into equation 7.15 the diffusion terms ∇N_{gb} caused by a gradient of concentration we obtain the following relationship for the flow:

$$J_{gb} = \frac{\delta D_{gb}}{d} \left(-\nabla N_{gb} + \frac{N_{gb}}{kT} Z_{gb}^* eE \right) \qquad (7.16)$$

The analysis of experimental data for grain boundary diffusion is more complicated than the analysis of data for lattice diffusion because of the variety of paths that can exist in that case. As grain boundary diffusion is usually accompanied by lattice diffusion, the parallel flow of diffusing atoms in the lattice adjacent to the boundary is inevitable. If there is a local net divergence in the atom flow matter will accumulate in some areas of the film while in other places vacancies appear. On reaching a critical concentration of vacancies, voids can nucleate in a film because of vacancy condensation. The formation of voids and cavities decreases the local cross-section of the film conductor and ultimately breaks the connection of the integrated circuits.

7.1.7 Effective charge of ions

For metals with electronic conductivity atoms migrate to the positive electrode. Experiments on electromigration in pure metals give the magnitude of the effective charge $q_{ef} = -(5-50)e$. Data on Z_{gb}^* and Z_i^* for films of various metals with electronic conductivity are given in Table 7.1 [7.2]. It can be seen that the effective charge for grain boundary motion hardly

Table 7.1 Data on the effective charges of ions in films for some metals [7.2]

Impurity in metal	Metal of film	T, K	Z^*_{gb}	Z^*_i
Cu	Al	528	−16.8	−7
Cu	Al + Cu	448	−20	−7
Mg	Al + Mg	504	−40	−39
Ni	Al + Ni	448	−3	−
Al	Al	500	−10	−17
Al	Al + Cu	448	−30	−17
Al	Al + Mg	504	−30	−17
Al	Al + Ni	448	−30	−17
Ag	Au	523	−9.6	−7.4
Au	Au	573	−10	−9.2
Au	Au + Ta	569	−	−
Sb	Ag	750	−230	−100

differs from the effective charge of a crystal lattice although the diffusion coefficients differ by many orders of magnitude. Thus, the observed difference of the electromigration on the lattice and on grains boundaries is connected with the difference in the mobility of the atoms. Therefore by reducing the mobility along grain boundaries, it is possible to reduce the damage produced by electromigration.

For lattice electromigration the experimental and theoretical values of Z^*_i agree well but there is not enough data about the magnitudes of Z^*_{gb} for the migration along the grain boundaries. For platinum the effective charge Z^* is about +0.3 [7.3]. In this metal the direction of the mass transport is determined by the direction of the electrostatic force. It is possible to determine the contribution to the effective charge produced by the electron wind force even for small values of the effective charge. The separation of the relative contribution of the electrical field and electron wind field depends on the value of Z_{ei} in equation 7.3. For metals having hole conductivity the absolute values of the forces F_z and F_e in equation 7.6 are added. Thus the effective charge will be positive. For tungsten it was found that $Z^* = +20$ and for iron $Z^* = +9.5$. The effect on the ions of holes in these metals has been confirmed experimentally [7.2, 7.3]. Thus the ions are displaced in the direction of the cathode.

The force due to the electron wind in equation 7.3 is proportional to the mean free path of an electron. As this decreases with an increase in the temperature, the effective charge Z^* will decrease accordingly. For metals with electron conductivity, electromigration experiments confirm a decrease in the effective charge of the host ions with an increase in the temperature although this effect can be weak. The experiments also show a decrease in the effective charge with an increase in temperature for impurity atoms. Thus, it was established for copper in aluminium that for an increase in temperature from 325 °C to 500 °C the effective charge decreased from −14.9 to −4.1. This charge was assumed to be due to transport along the grain boundaries [7.2].

7.1.8 Divergence of the ion flow

Tests on thin-film conductors have shown that electromigration damage, as a rule, does not occur at the ends of a conductor (at the cathode or anode) but is distributed randomly on different sites along its length. To initiate electromigration damage at any particular part of a film it is necessary to have a net divergence of the ion flow. Thus damage to a metal film arises in areas of local heterogeneity. Let us consider the factors that may create a divergence in the flow by an analysis of the continuity equation, in which for simplicity we will consider only grain boundary transport. For polycrystalline materials, which covers most thin-film conductors, the total flow of atoms is equal to the sum of the flows on the crystal lattice and along the grain boundaries, given by equations 7.14 and 7.15. However for the majority of thin films, the mass transport along the grain boundaries usually dominates.

In pure metals the vacancy mechanism of diffusion usually plays a basic role. Here the atom jumps from a full lattice site to an empty lattice site, an adjacent vacancy, after a local fluctuation of the energy. Thus it is possible to assume that the concentration of activated atoms is equal to the concentration of vacancies. The vacancies occur mainly on the grain boundaries.

Let us write the continuity equation for the vacancy transport of atoms along the grain boundary:

$$\frac{\mathrm{d}N_\mathrm{v}}{\mathrm{d}t} = -\nabla J_\mathrm{v} + \frac{N_\mathrm{v} - N_0}{\tau} \tag{7.17}$$

where N_v and N_0 are the concentrations of the excess and equilibrium vacancies on the grain boundaries, τ is an average time life of the excess vacancies due to sinks and sources and J_v is the flow of vacancies.

From equation 7.17 it follows that the local concentration of vacancies depends on the divergence of the flow and the relaxation time of surplus vacancies. The steady-state condition, $\mathrm{d}N_\mathrm{v}/\mathrm{d}t = 0$, occurs when the excess number of vacancies, $N_\mathrm{v} - N_0$, becomes proportional to the flow divergence. In this steady-state condition the vacancy flow into any volume is equal to the number of annihilating vacancies.

The expression for the flow of vacancies, allowing for the diffusion term which is proportional to the gradient of the vacancy concentration, is by analogy with equation 7.16:

$$J_\mathrm{v} = \frac{\delta}{d} D \left(-\nabla N_\mathrm{v} + \frac{N_\mathrm{v}}{kT} Z_{\mathrm{gb}}^* eE \right) \tag{7.18}$$

where D is the diffusion coefficient of the vacancies.

The second term in this expression is the contribution from electromigration. Calculation of the flow divergence, ∇J_v, allowing for the dependence of the film parameters N_v, Z^*_{gb}, T, d, D on the co-ordinates and the insertion of their values into expression 7.17 reduces to the following complicated equation [7.2] in which practically all factors that can result in a divergence of the atom flow are included:

$$
\frac{dN_v}{dt} = \frac{\delta}{d} D \left[\nabla^2 N_v - \frac{N_v}{kT} Z^*_{gb} eE \left(\frac{\nabla N_v}{N_v} + \frac{\nabla Z^*_{gb}}{Z^*_{gb}} - \frac{\Delta T}{T} \right) \right]
$$
$$
- \left(\frac{\nabla D}{D} - \frac{\nabla d}{d} \right) \frac{\delta}{d} D \left(-\nabla N_v + \frac{N_v}{kT} Z^*_{gb} eE \right) + \frac{N_v - N_0}{\tau}
$$

(7.19)

Diffusion can also occur due to an external force produced by a mechanical stress gradient. To allow for any external forces directed parallel to the grain boundary in equation 7.19 it is necessary to add a term $\frac{\delta}{d} D \frac{F_{gb}}{kT} \nabla N_v$ where F_{gb} is an external force acting on the grain boundaries in the direction of the electron flow.

The net divergence of the atom flow can be caused by gradients in many quantities: the defect concentration, temperature, effective charge, mechanical stress, diffusion coefficient and grain size. The last three factors are specific to thin films and in bulk samples they are usually neglected. In thin films the divergence of the flow can be equal to zero only if there is mutual compensation of all the items in equation 7.19. This is extremely unlikely.

Almost all the factors of heterogeneity of the atom flow in equation 7.19 are connected with the microstructure of the film. The exception is the gradient of temperature. The temperature gradient in metal films is usually insignificant, as they have good thermal contact with the substrate. Nevertheless, as theory and experiment shows, it is impossible to neglect the effect of local Joule heating.

Thus the random distribution of electromigration damage along the film length is explained by the random distribution of the heterogeneity. This eliminates an explanation of electromigration solely because of the temperature gradient and shows the possibility of the simultaneous origin of the positive and negative divergence of the flow in a small length of the film. From equation 7.19 one can see that the structural heterogeneity of a film is determined by the terms:

$$
- \left(\frac{\nabla D}{D} - \frac{\nabla d}{d} \right) J_v \quad \text{and} \quad \frac{N_v}{kT} Z^*_{gb} eE \left(\frac{\nabla N_v}{N_v} + \frac{\nabla Z^*_{gb}}{Z^*_{gb}} \right)
$$

The modification of a structure is the basic reason for the spatial dependence of the vacancy concentration in a film. Let us write the well-known expression for the diffusion coefficient of vacancies:

$$D = D_0 \exp(-E_a/kT) \tag{7.20}$$

where D_0 is the exponent prefactor and E_a is the activation energy of diffusion.

After differentiation we obtain a formula for the gradient of the diffusion coefficient:

$$\nabla D = \frac{D_0}{kT}\left(-\nabla E_a + \frac{E_a}{T}\nabla T\right) \tag{7.21}$$

In this last expression we have neglected the term ∇D_0. The simultaneous calculation of all the factors causing a divergence in the ion flow in equation 7.19 is difficult. Therefore the analysis of the equation is usually made for each factor separately. The influence of the various structural factors is considered in the following section.

7.1.9 Influence of the structural heterogeneity on electromigration

In typical films the heterogeneity of the crystal structure is such that the damage appears in the sections of the film near its ends. Experiments have shown the dependence of the electromigration rate on the structure of the film. This has been detected with the help of electron microscopy. Voids arise most often at the junction of the grain boundaries at so-called "triple points". This effect is most obvious in fine-grained films with a high density of grain boundaries and it is less frequent in large grain films. It is possible to conclude that an increase in the grain size should produce an increase in the median time to failure for thin-film conductors.

Figure 7.4. shows the dependencies of the median time to failure on temperature for two Al films with a thickness of 1.2 µm, but with the different grain sizes [7.7]. The films were made by electron-beam evaporation in vacuum (the pressure of the residual gases was 1.33×10^{-5} Pa) on substrates of oxidised silicon at a condensation temperature of $T_c = 200\,°C$ and 500 °C for the large and small grain films. The films that were deposited at a temperature of 500 °C had a large grain structure with average grain size of $d = 7.8$ µm and dominant crystal orientation (111) (98%). The films made at 200 °C had a small-grained structure ($d = 1.2$ µm) with mixed orientation (111) and (110). The average median time to failure for the large grain films was $t_{50} = 8127$ hours with a variance $\sigma = 0.29$. This was found to be four times more than for the films with the small grains. For the latter $t_{50} = 1995$ hours with $\sigma = 0.47$. The activation energy of electromigration for the large grain films was $E_a = 0.73$ eV and for fine-grained films $E_a = 0.51$ eV. The higher activation energy for the large grain film can be partially explained by an increase in the contribution of a crystal lattice to the transport of atoms. The increase in the activation energy with an increase in the grain size can also be explained by the appearance of a dominant orientation (texture) which is more strongly expressed in the large grain films than in the

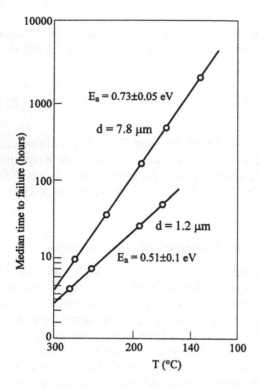

Figure 7.4 Dependence of the median time to failure on the inverse temperature for two Al films with a thickness of 1.2 μm with the different
grain sizes [7.7] for a large grain film ($d = 7.8$ μm) and for a fine
grain film ($d = 1.2$ μm).

fine-grained films. The latter always have a more chaotic orientation. In the
fine-grained films there are many paths for the diffusion.

Figure 7.5 shows a schematic diagram of a triple point (the junction of the
three grain boundaries 1, 2, 3) with angles θ_1, θ_2, θ_3 between the grain
boundaries and the direction of the net flow of the charge carriers along
the x-axis. The diffusion coefficients along the grain boundary are taken to
be different for the three boundaries. The width of all three grain boundaries
is taken to be equal to δ. For there to be no depletion or accumulation of
matter in the region of the triple point, the flow of atoms leaving the region
of the triple point should be counterbalanced by the incoming flow. The
resulting flow from equation 7.18 but disregarding the diffusion term ∇N_v
can be written as [7.3]:

$$J_v = \frac{\delta}{d} \frac{N_v Z_{gb}^* eE}{kT} (D_1 \cos \theta_1 - D_2 \cos \theta_2 - D_3 \cos \theta_3) \tag{7.22}$$

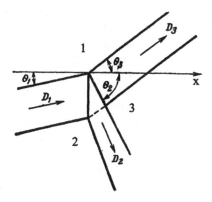

Figure 7.5 Schematic diagram of a triple point with angles θ_1, θ_2, θ_3 between the grains boundaries (1, 2, 3) and the direction of the net charge carriers flow J_e along the x-axis.

The term $D_1 \cos\theta_1$ expresses the flow coming into the triple point and the terms $(D_2 \cos\theta_2 + D_3 \cos\theta_3)$ give the leaving flow. Which is the larger contribution depends on the condition of boundaries. Consider, for example, that the grains are oriented randomly. Then we may expect that $D_1 \neq D_2 \neq D_3$. Because of the large variation in the properties and diffusion coefficients of the grain boundaries it is possible that the flow on one of the boundaries will be more than on the other two adjacent boundaries. The inequality of the diffusion coefficients on the grain boundaries determines the rate of growth of the damage. Fast migration along one of the boundaries (with diffusion coefficients D_2 or D_3 in Figure 7.5) will give a mass inflow from the right of the triple point and outflow from the left, that results in damage at the triple point.

The rate of mass transport is connected with the orientation of the grains. The diffusion along the grain boundary is anisotropic. Thus the diffusion parallel to the boundaries happens with a rate some orders of magnitude greater than in a direction perpendicular to the boundaries. A film with large grains in a bar structure can be considered as an ideal model in which the grain boundary is represented by a great number of parallel dislocations oriented perpendicularly to the substrate and the direction of mass transport. The velocity of mass transport in such metal films is lower than in fine-grained films with randomly oriented crystals, where the grain boundaries are not regular and the component of the diffusion velocity along the direction of the current is large.

It has been established that the activation energy of electromigration in films with texture is rather higher than in films with random orientation of grains. In textured films the damage is considerably reduced because of the existence of the triple points. However, in textured films another structural heterogeneity begins to play a role, namely the mixture of various grain sizes. In films deposited at rather high temperature, in which the nominal size of the grains exceeds 3–4 μm, a wide distribution of grain sizes can be observed.

The reason for the increase in damage in films containing grains with a range of sizes is clear from Figure 7.6. The damage in this case occurs because of the sharp change in the number of diffusion paths in the direction of the electron flow, rather than the change in the properties of the separate boundaries. The transport of matter occurs more strongly in those places where there is a denser network of grain boundaries, that is, in those places with a smaller sized crystal structure. If from some cross-section of a film there are more boundaries removing matter than depositing, then there is a tendency for accumulation of vacancies and, therefore, the formation of hollows and voids. Figure 7.6. illustrates diagrammatically the case when one grain boundary (2′) feeds into the cross-section ABC while grain boundaries 5, 6, 7, 8 and 9 remove material. In this case a depletion of matter will occur behind the cross-section ABC, where voids will form. Also matter will accumulate as growths in front of ABC because of the atoms diffusing on grain boundaries 1, 2, 3 and 4.

It should be noticed that significant damage can arise in films with a broad distribution of grain sizes in the case when individual grains cross the whole width of the film. It has been shown that in large grain aluminium films about 80% of the failures occur because of structural heterogeneity of a similar type [7.3]. A divergence of the atom flow can also occur near atoms of impurities, particles of metal oxide, micro-voids, etc.

The dependence of electromigration damage on the structural heterogeneity is shown by the extremely diverse statistics of the lifetime data of conductors in real use. With a decrease in the character and number of the

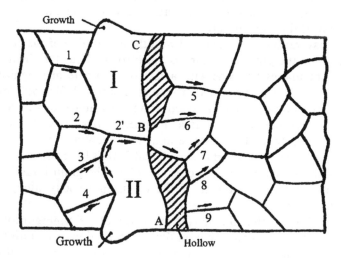

Figure 7.6 The increase of damage in a film containing grains of various sizes.

various types of damage mechanisms in films the magnitude of the standard deviation in the distribution of the time to failure (equation 7.2) decreases, independent of the median time to failure. This means an increase in the time to the first failure, which is the one that is of most interest for the reliability of real commercial devices. For example, for a textured film it is likely that a failure caused by the presence of a triple point is less probable.

The experimental values of the standard deviation for textured films with (111) orientation appeared less than for films with a random structure ($\sigma_f = 0.29$ and 0.47 respectively). Thus, the ordered orientation of crystals in the films, as well as the increase in the activation energy and average time to failure, decreases the standard deviation so that the failure rate is better defined.

7.1.10 Electromigration in multicomponent systems

Tests of the electromigration resistance of aluminium thin-film conductors have shown that by adding copper at 4%wt. the median time to failure increased approximately 70 times for the same current density, $j = 4 \times 10^6$ A cm^{-2}, and temperature, 175°C. Damage caused by structural heterogeneities were distributed randomly along the conductors. The median times to failure were approximately 6 and 400 hours. The study of the relation between the concentration of copper in aluminium and the service life shows that at low concentration of copper the lifetime increases monotonically with the increase in the copper content up to a concentration of about 3–4%wt. However at higher concentrations of copper this relation fails [7.3].

It has been found that most substances dissolved in aluminium (Ag, Au, Cu) increase the electromigration in the bulk crystal. Therefore the positive influence of impurities in thin Al films on the electromigration resistance cannot be because of lattice processes.

Further studies of the mass transport in thin films of the Al/Cu alloy have shown that the increase in the lifetime is connected with a decrease in the grain boundary diffusion when there is a minimum content of impurity in the conductor and it is mainly concentrated in the grains boundaries.

In Table 7.2 [7.2] data are given on the median time to failure for conductors of Al, Au and Cu with the addition of other metals and silicon. These results were obtained under comparable conditions for the films' manufacture and identical conditions of test, or by an extrapolation of equation 7.1 using values of n and E_a from the literature.

To evaluate the average service life of conductors it is necessary to realise that the predictions can vary considerably because of the sensitivity of the experiments on the test specifications, geometries and the microstructures of the actual specimens. Thus, the values given in Table 7.2 for the median time to failure give only rough estimates.

The explanation of the electromigration resistance of these alloy films is based on the relative energy of the impurity atoms at the grain boundaries.

Table 7.2. The median time to failure for conductors of Al, Au and Cu with the addition of other metals ($j = 2 \times 10^6$ A cm^{-2}) [7.2]

Host metal	Impurity, %weight	t_{50}
Al at 175 °C	–	30–45
	Si 1.8	100–200
	Cu 4	2500
	Cu 4, Si 1.7	4000
	Ni 1	3000
	Cr	8300
	Mg 2	1000
	Cu 4, Mg 2	10000
	Cu 4, Ni 2, Mg 1.5	32000
	Au 2	45–55
Au at 300 °C	Ta	4000
	Ni-Fe	800
	Mo	100
	W + Ti	25
Cu at 300 °C	–	180
	Al 1	300
	Al 10	6000
	Be 1.7	20000

Figure 7.7 shows schematically the segregation of the atoms of the solute on the boundary imperfections (steps and dislocations). The amount of the dissolved substance on the grain boundaries is determined by the binding energy, E_{NS}, of the impurity atom with the imperfection. At large enough binding energy the number of lattice sites on the grain boundary suitable for diffusion is strongly reduced. For example, for $E_{NS} = 0.2$ eV the presence of 1% of impurity atoms (at a temperature of $T = 400$ K) decreases the number of unfilled lattice sites (vacancies) by more than one order of magnitude. This decrease in the number of lattice sites that are available for diffusion results in a drop of the mass transport.

If the diffusion is controlled by the jumps of atoms of the host metal between vacancies, then the diffusion will be controlled by the presence of impurity atoms on the dislocations. Then the dissociation of a vacancy-impurity atom complex is necessary. That is the activation energy of the diffusion process is increased if there are impurity atoms on the grain boundary. The experimentally determined activation energy for films of Al/Cu alloy appeared higher by approximately 0.2 eV than for films of pure Al. This magnitude apparently corresponds to the binding energy of a Cu atom with the grain boundary.

The solubility limit of Cu in Al is less than 1% at all operating temperatures so that during annealing the excess copper atoms form inclusions of Al$_2$Cu. These increase the heterogeneity. The copper atoms electromigrate

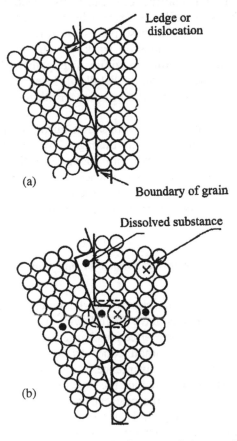

Figure 7.7 Schematic diagram of the step imperfections on the grain boundaries of a pure metal and the segregation of the dissolved atoms on
the boundary imperfections [7.3]: (a) pure metal; (b) segregation of
the dissolved substance atoms on the grain boundaries and other
imperfections.

slowly along the grain boundaries but are replaced by copper from the
inclusions. When all the excess copper has migrated to stable places the
grain boundaries are empty and aluminium can migrate more freely to result
in rapid failure.

7.1.11 Fabrication imperfections which accelerate electromigration

Let us now consider which fabrication defects in integrated circuit interconnects can enhance electromigration and create the potential for IC failure.
Such defects may be: heterogeneity of the film, non-uniformity of the

thickness, the presence of differential contractions and steps, mechanical damage, etc. As a rule, the thickness of a film decreases where it is deposited over steps on the insulating oxide SiO_2. In these places the formation of micro-cracks and voids is possible. Such steps on tracks which carry the greatest current stress are especially susceptible to failure. The quality of the film deposition over the oxide steps depends on the details of the equipment and methods used. Usually the surface is made flat in a planarisation process before the film is deposited.

It is difficult to detect imperfections in the metallisation during fabrication. For example, if the metallisation is partially damaged, by a scratch, crack or particle, the parameters of the IC will not change significantly. However during storage, and especially during use when there is current stress, the gradual transport of metal occurs because of electromigration and later a full rupture of the film will take place in the region of the initial damage.

7.1.12   Methods to increase electromigration resistance in thin-film conductors

Experiment, and related computer modelling, shows that the material which is resistant to electromigration damage must have the following properties: a high activation energy of self-diffusion; a low effective charge for ions in an electric field; a low diffusion coefficient, small concentration gradients, homogeneous sized grains to eliminate flow divergence of ions near anomalously large grains; large grain structure to reduce the flow of atoms along grain boundaries; consistent orientation of the grain structure (texture) of the film to decrease the diffusion along the grain boundaries and to decrease the diffusion coefficient gradient; lack of heterogeneous inclusions such as impurities, metal oxide or imperfections which could act as centres for vacancy condensation with the subsequent formation of voids and hollows. In this design there should be low electric current density and no temperature gradients along the film. The temperature gradients can be reduced by the appropriate design of the track sizes for the expected current flow and by good cooling. Let us consider in more detail the factors which define the electromigration immunity and the methods which reduce the rate of mass transport in thin film conductors.

The activation energy of electromigration

If we consider that the mass transport in films during electromigration is determined by diffusion, then the activation energy of electromigration should be equal to the activation energy of self-diffusion, which in turn is connected with the melting temperature of the metal. The direct proportionality between the activation energy and the melting temperature has been established experimentally for metals (Figure 7.8.) [7.8]. The data for the

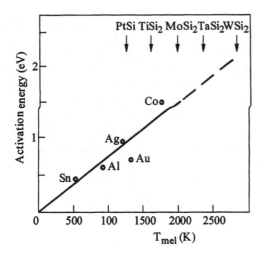

Figure 7.8 Dependence of the electromigration activation energy in different metals on their melting temperature [7.8].

metal films correspond to the activation energy of electromigration along grain boundaries. At higher temperatures the straight line is approximated by the dotted line. Here the possible activation energy for silicides can be deduced from the melting temperatures shown.

A statistical analysis of data for many metals shows a simple correlation between the activation energy of self-diffusion on a crystal lattice and the melting temperature:

$$E_a = 18kT_{mel} = 1.55 \times 10^{-3} T_{mel}(eV) \tag{7.23}$$

where $k = 8.6 \times 10^{-5} \, eV \, K^{-1}$ is the Bolzmann constant and T_{mel} is the melting temperature.

Films of the refractory metals, Mo and W, have higher values of activation energy for electromigration. The application of these materials for thin-film conducting layers should increase the service life compared with conductors made from Al, Cu, Ag and Au. However, they have a high resistivity which increases the delay in the interconnect. Also they are difficult to make because of the photo-lithography and the high annealing temperature. The refractory metals, Mo and W, oxidise at low temperatures to form volatile oxides. This can result in the destruction of the film. Now the most common metals used for the IC interconnection is Al, in particular alloys of Al/Cu, Cu and its alloys and silicides of the refractory metals. Copper is becoming the dominant metal for the interconnects in high performance ICs with fine resolution. This is because copper has a lower resistivity than aluminium, although it is more difficult to process.

Size of the grains

An increase in the grain size is the most direct way to increase the lifetime of thin-film conductors. The high activation energy in large grained films can also be controlled by the texture, which is more uniform in large grain films than in the fine-grained ones. If large grains completely fill the cross-section of the conductor then the structure is called "bamboo", which has few grain boundaries and then these are normal to the electric field so that there is no grain boundary electromigration. The lifetime increases greatly. However if there are even short regions of smaller grains and triple points there can still be a source of ion flow divergence. This results in a long median lifetime but high dispersion and, most important, only a marginal improvement in the time to first failure in a batch. For Al films with a thickness of 1 μm the largest size of the grain is 8–10 μm. If the condensation temperature is less than 400 °C, the grain size does not exceed 2–3 μm.

Dielectric overlayers

The deposition of a hard dielectric cover over thin film conductors increases the median time to failure due to electromigration. In these experiments the Al conductors are covered with a layer of various glasses or of Al_2O_3, obtained by anodic oxidation. The activation energy of electromigration is increased.

For the protection of the Al film surface it is possible to use silicon dioxide made by the decomposition of SiH_4 in oxygen or in the plasma of a high frequency discharge. The increase in the reliability of the thin-film conductor is explained by the formation of a metal–dielectric boundary with chemical bonding, which ensures strong adhesion between the oxide and metal. The boundary of the metal–dielectric acts as a poor sink for vacancies and the oxide is an impenetrable diffusion barrier, which would reduce excess vacancies. This reduces the mass transport on the film surface. Also a hard encapsulation resists the formation of extrusions or hillocks since a local high pressure region is formed and the resulting stress gradients produce a back flow for the electromigration. A similar effect reduces the elecromigration in very short lengths of line, for example between a contact and a via made from a different metal with a low diffusion. The electromigration forces the atoms against the barrier to atom flow so that a stress gradient builds up to resist the electromigration flow. This process is only effective if the local pressure does not exceed the elastic limit. This "Blech length" effect means that there is negligible electromigration if $(jl) \sim 4 \times 10^5 \, A \, m^{-1}$ for aluminium. The use of protective dielectric encapsulation is not effective in all cases. Some experiments have not detected any effect on the electromigration lifetime for different combinations of the conductors and dielectric encapsulation.

Application of alloys and other materials

It has already been mentioned in Chapter 1 that alloying of aluminium with some elements is an effective way to increase the electromigration immunity of films. It has been observed [7.9] that aluminium films containing 1% of the rare-earth metals holmium (Ho) or europium (Eu) have good properties for the manufacture of very large scale integrated circuits (VLSIs) with sub-micrometer size interconnections. The introduction of these elements produces a modification of the crystal structure of the thin aluminium films. Firstly, the grain size increases for a small alloy concentration and this decreases the electron scattering of the grain boundaries, which in turn causes a drop in the film's resistivity. Secondly, the rare earth alloys promote textured grains. The axis direction depends on the concentration of the alloy. Also the rare earths interact with the residual gases so that the film is cleared of undesirable impurities so that there is an additional decrease in the resistivity and improvement in the metal quality. Also the addition of alloys of special impurities such as yttrium up to 1% increases the re-crystallisation temperature.

The longest life has been obtained for thin-film conductors of the four-component alloy Al/Cu/Mg/Ni (see Table 7.2). However, in practice the use of this alloy is inconvenient. The introduction into Al films of a small amount of oxygen also increases the service life of a conductor by approximately 10 times. This is explained by a decrease in the grain boundary diffusion because of the stabilisation of the microscopic defects by the segregation of oxygen atoms on to the grain boundaries to be chemically bonded. Other ways of increasing the electromigration resistance have been reported in the literature. For example, the application of a Ti–Al layer structure.

Local self-heating of tracks can increase the grain size because of the growth of the energetically favourable grains at the expense of others during re-crystallisation (see Section 1.5.3). In these conditions the rupture or delamination of the metal can occur because of the large tensile stresses produced at the places where grains grow. The operation of these failure mechanisms can be decreased at the expense of a drop in the current density flowing along the tracks.

7.2 Corrosion and oxidation of metal interconnects

The destruction of thin-film metallisation can take place by chemical or electrochemical (electrolytic) corrosion. Corrosion of the aluminium metallisation occurs due to the contamination of the crystal surface by various reactants and water. Moisture can penetrate the case of an IC if the encapsulation is broken. Aluminium reacts with water inside the circuit encapsulation in the presence of ions of chlorine, copper, iron, ammonium and other chemical elements. The corrosion destroys the aluminium film.

Such elements as chlorine can remain on a substrate after plasma and ion etching. They react with the aluminium in the presence of moisture by the following two reactions even without the application of an electric field [7.4]:

$$Al + 3HCl = AlCl_3 + \frac{3}{2}H_2 \tag{1}$$

$$AlCl_3 + 3H_2O = Al(OH)_3 + 3HCl \tag{2}$$

The chlorine corrosion process is not exhausted after the formation of $Al(OH)_3$ since it promotes further reactions with the unprotected aluminium. The aluminium track is gradually covered with white flakes of aluminium hydroxide and the metal breaks away first at the place of contact, near a positive potential. The rate of corrosion depends largely on the voltage applied to the circuit. A potential difference of 5 V is sufficient for rapid corrosion to take place [7.1]. The rate of corrosion depends on the distance between the electrodes, the ambient temperature and the concentration of the impurity ions on the surface of the crystal. An analysis of failures from corrosion shows that they develop first on grain boundaries with the formation of continuous micro-cracks, which produce a breakdown of the metallisation. The process is accelerated by an increase in the temperature. Aluminium corrosion occurs faster in those places where the current density is higher, for example in regions where there are imperfections in the metallisation such as a step or scratch, and also in places with a contact potential difference such as a p–n junction. Where the tracks are close together the corrosion can produce short-circuits between conductors. The metal electrode under the positive potential is transferred gradually to the negative electrode to form a conducting track.

Chemical reactions can produce both open circuit of the track and short-circuits between the neighbouring tracks. The rate of corrosion can be increased many times if there are tensile mechanical stresses.

The encapsulation by phosphorous-silicon glass with an excess of phosphorus can considerably increase the corrosion since the surplus phosphorus interacts with any water to form phosphoric acid, which increases the corrosion of the metal. A decrease in the concentration of phosphorus in the phosphorous-silicon glass covering the aluminium metal by up to 5%wt. increases the median time to failure by corrosion by more than three orders of magnitude.

A basic defect of a component which may enable a chemical or electrochemical reaction is the non-hermeticity of the case. The penetration of moisture through a leak in the case, adsorption on the surface of the metal and then through pores and cracks in the protective coatings on the surface of the crystal produces metal corrosion, which frequently has an electrochemical character if there is ionic contamination.

In parallel with electromigration and electrochemical corrosion there is a degradation mechanism of metal films connected with the oxidation of aluminium that leads to an increase in the track resistance [7.1]. The mechanism of failure in this case consists of the interaction of the oxygen with the surface of the grains in the volume of the film material. The growth of an oxide film on the surface of the conducting track and the formation of oxide inter-grain layers decreases the effective cross-section of the conductor and, consequently, the track resistance and the resistivity of the material film are increased. With an increase in the local resistivity, reduction in the local cross-section or delamination from the substrate heat sink, the local temperature of the metal increases to produce an increase in the rate of the electromigration and failure.

7.3 The $1/f^2$ noise in metal films

The relationship between the properties of the $1/f^\gamma$ noise and the rate and amount of electromigration damage of thin metal films has been studied extensively. The results are varied because of the variation in the samples but can be summarised. At low currents the noise has a good $1/f$ spectrum and the intensity rises as j^2 as expected from equilibrium resistance fluctuations. The mechanism is the equilibrium fluctuation in the number of vacancies and the motion of the vacancies, or atoms, described in Chapter 5.

It has been demonstrated that the electromigration component of $1/f$ noise arises from the diffusive motion of atoms mainly along the grain boundaries. Indeed, the activation energy determined using the temperature dependence of the noise PSD corresponds to the activation energy of diffusion along the grain boundaries. This was confirmed in experiments [7.10] in which the activation energy of $1/f$ noise in Al films and its alloys increased together with a rise in the activation energy of diffusion along the grain boundaries with the growing concentration of impurities. Since impurities segregate at the grain boundaries, they do not appreciably influence the bulk properties of the crystals.

In submicron metal conductors with a bamboo structure, so that there are no grain boundary paths for motion along the current direction, the activation energy determined from the temperature dependence of the PSD of the $1/f$ noise is equal to the activation energy of diffusion through the bulk crystal lattice (1.45 eV). In some samples a lower energy has been found corresponding to motion along the interface between the metal and its surroundings.

At high current densities, $j \sim 10^6$–$10^7 \mathrm{A\ cm}^{-2}$, and at elevated temperatures when significant electromigration takes place, the current dependence of the noise PSD differs from quadratic, and $S \propto j^a$, where the observed values were $a = 3$–4 and even 7 [7.11]. Also, the frequency exponent, γ, increases with increasing current. As another example, a in Ni–Cr films grew from $a \approx 2$ at weak currents up to 3.5–4.5 at larger ones (similar to Figure 6.5), while the frequency exponent, γ, rose from 1 to 2.5 [7.12].

The deviation of the current dependence of the electromigration noise from quadratic indicates that this component of flicker noise originates from non-equilibrium fluctuations of the conductivity.

A detailed study of the noise in Al/Cu samples with different direct current and temperature stress has shown that the current and frequency exponents increase as the stress increases and also as the samples become more damaged [7.13]. This is shown in Figure 7.9. The $1/f$ noise in metals is small so that very low frequency measurements are made. The voltage signal is acquired over a period of time and the Fourier transform calculated to obtain the spectrum. Normally the time signal itself is not recorded or studied. In these experiments the noise was measured directly using a spectrum analyser but also the resistance changes were recorded at two-second intervals. This gave a low frequency spectrum which joined the other. The measurements shown in Figure 7.10 reveal that during electromigration, especially on a sample which has suffered damage due to previous electromigration, the resistance fluctuations are highly non-equilibrium and show upward and downward steps. It is these that produce the changes in the noise behaviour.

These events have been shown, by electron microscopy during stress, to be due to the motion of voids past obstructions, void creation, growth and healing. This will be discussed further in Section 7.5.

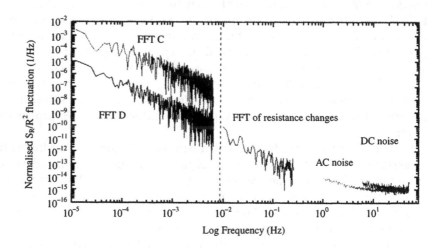

Figure 7.9 The normalised resistance noise spectra for damaged samples P2D4 and P2D6(2). The high frequency data are for sample P2D4 stressed at $J_{DC} = 2.23 \times 10^6\,\mathrm{A\,cm^{-2}}$ and 250 °C. The low frequency data are for sample P2D6(2) stressed at $J_{DC} = 2.51 \times 10^6\,\mathrm{A\,cm^{-2}}$ at 250 °C. The frequency exponents are 1.7–1.9, 1.4–1.6 and 2.0–2.2 for DC, AC and FFT methods for sample P2D4 and 1.8 and 2.1–2.3 for FFT C (upper) and FFT D (lower) for sample P2D6(2) [7.13].

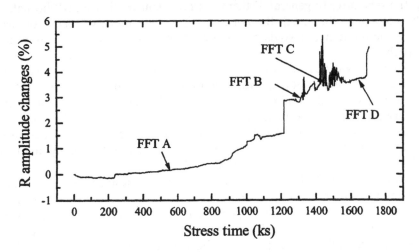

Figure 7.10 The resistance variation during the full lifetime of sample P2D6(2) stressed with $J = 2.51 \times 10^6\,A\,cm^{-2}$ at 256 °C. The various time periods shown are used to calculate the FFT spectra shown in Figure 7.9 [7.13].

7.4 Fast reliability tests

The reliability, or lifetime under normal use, is an important commercial quantity for electronic components. Such assessment will naturally take a very long time so that accelerated test methods are used with large temperature, voltage or current stresses and an extrapolation is made of the results to estimate the life under normal operating conditions. The extrapolation has to assume that the same degradation processes are taking place at the high and low temperatures. A valuable quantity for measurement would be a 'reliability indicator' which showed a large change at early stages of degradation and would predict failure. Such a quantity would also be able to assess the quality of a device after manufacture to identify poor quality devices, or the best devices to select for critical applications, and be used to assess changes to the process variables during manufacture.

The slow change of the static parameters of a device are used as reliability indicators. Examples are the change in the resistance or threshold voltage. It is found that the excess, $1/f$ noise, is a very sensitive reliability indicator for many electronic devices with perhaps an order of magnitude change in size whilst the static parameter may change by only a few percent [7.11, 7.14]. A characteristic of excess noise is that it can arise from many sources but these are mainly related to defects. Also there may not be any direct link with lifetime to failure.

Specifically for thin metal films there is considerable evidence that the noise increases if a film has been damaged by electromigration to leave voids in the structure. In general, if there is a noise source distributed throughout the volume of the metal, the noise across the terminals will increase faster than the resistance if voids are created to produce a non-uniform current distribution.

There are various tests that have been developed for assessing the electromigration resistance of a specific technology. These are destructive, but can be performed on the wafer for rapid results. Normally special test structures are used to reduce effects due to the special conditions. Since the stress is so high any extrapolation of the results to normal conditions or to real interconnects must be treated with caution. The methods usually used are TRACE, BEM, WIJET and SWEAT [7.5].

Menon and Fu [7.15] proposed a fast, although destructive, test to detect defects in samples. The resistance is measured as the current is increased steadily until the sample goes open circuit. For a resistance with a positive temperature coefficient and driven by a current source there is positive thermal feedback so that the sample heats to the melting point at a critical current J_C given by:

$$J_C^2 = (\alpha R_T R_0)^{-1} \tag{7.24}$$

where R_0 is the initial resistance, α is the temperature coefficient and R_T is the thermal resistance to the substrate. The resistance should then vary as:

$$\frac{R_0}{R} = 1 - \left(\frac{J}{J_c}\right) \tag{7.25}$$

Deviations from the straight line can be ascribed to weak spots in the line where the fusing takes place.

7.5   Second harmonic generation in electromigration damage

A specific method of generation of second harmonic signals is found in metal thin-film samples. An alternating current generates Joule heating at twice the frequency. This temperature rise gives a resistance increase through the temperature coefficient of resistance. This second harmonic change will appear as a voltage change if a direct current is also passed.

Provided the thermal time constant for heat flow to the surroundings is long compared with the signal period, the second harmonic voltage is given by [7.16]:

$$\frac{3}{2} I_{DC} I_0^2 R_0 R(t) \alpha \beta \tag{7.26}$$

where the current has a DC component and an AC component.

For a uniform thin-film sample this can give information about the value of the thermal resistance to the substrate and, from the frequency dependence, the thermal time constant. If the degradation is uniform, on a scale of the line width and thickness, the changes in the resistance and second harmonic give the average changes to the resistivity and the cross-sectional area. For non-uniform thin films, such as samples with voids created by electromigration damage, the changes to the local resistivity and cross-section can be found if any sharp change can be ascribed to a very local heating [7.16].

This analysis can be very valuable to detect delamination since then the thermal resistance increases. Also the phase of the second harmonic may change if the thermal time constant and the signal period are comparable in magnitude.

Literature for Chapter 7

7.1   Chernishov, A.A., *Bases of the Reliability of the Semiconductor Devices and Integrated Micro-circuits* (in Russian), Radio i Svyaz', Moscow, 1988.

7.2   *Thin Films-Interdiffusion and Reactions*, Poate, J.M., Tu, K.N. and Mayer, J.W. (eds), J. Wiley, 1978.

7.3   D'Heurle, F. and Rosenberg, R., "Electromigration in thin films", *Physics of Thin Films*, Hass, G., Francombe, M.H. and Hoffman, R.W. (eds), Academic Press, New York and London, 7 (1973) 243–303.

7.4   *VLSI Technology* (ed.) Sze, S.M., McGraw-Hill, New York, 1983.

7.5   Scorzoni, A., Neri, B., Caprile, C. and Fantini, F., "Electromigration in thin-film interconnection lines: models, methods and results", *Materials Science Reports*, 7 (1991) 143–220.

7.6   Fiks, V.B., "On the mechanism of the mobility of ions in metals", *Soviet Phys.-Solid State*, 1 (1959) 14.

7.7   Attardo, M.J. and Rosenberg, R., "Electromigration damage in aluminium film conductors", *J. Appl. Phys.*, 42 (1971) 2381–2386.

7.8   Murarka, S.P., *Silicides for VLSI Applications*, Academic Press, New York and London, 1983.

7.9   Koleshko, V.M. and Belitsky, V.F., "Physical effects in metal films of submicrometer sizes and ways of the reliability and longevity raising for the VLSI interconnections" (in Russian) Vestnik AN BSSR. *Series Physical and Mathematics*, 66 (1982) 99–107.

7.10 Koch, R.H., Lloyd, J.R. and Cronin, J., "1/f noise and grainboundary diffusion in aluminum and aluminum alloys", *Phys. Rev. Lett.*, 55 (1985) 2487–2490.

7.11 Jones, B.K., "Electrical noise as a measure of quality and reliability in electronic devices", *Advances in Electronics and Electron Physics*, 87 (1994) 201–257.

7.12 Touboul, A., Verdier, F. and Herve, Y., "Lf noise measurements applied to the determination of electromigration mechanisms in Al (Si, Ti) and Ni-Cr stripes", *Proc. Int. Conf. Noise in Physical Systems and 1/f Fluctuations*, 1991, Kyoto, 73–77.

7.13   Guo, J., Jones, B.K. and Trefan, G., "The excess noise in integrated circuit interconnects before and after electromigration damage", *Microelectron. Reliability*, **39** (1999) 1677–1690.

7.14   Jones, B.K., Dorey, A.P., Richardson, A.M.D. and Xu, Y.Z., *Rapid Reliability Assessment of VLSICs*, Plenum Press, New York, 1990.

7.15   Menon, S.S. and Fu, K.Y., "A fast wafer-level screening test for VLSI metallisation", *IEEE Electron. Dev. Lett.*, **14** (1993) 307–309.

7.16   Jones, B.K., Guo, J., Xu, Y.Z. and Trefan, G., "Evolution of the electrical properties of interconnects under electromigration stress", *Mat. Res. Soc. Symp. Proc.*, **516** (1998) 9–13.

Index

Milton Keynes UK
Ingram Content Group UK Ltd.
UKHW020029071024
449327UK00032B/2983

9 780367 395131